TAKE A TECHNOWALK

To Learn About Mechanisms and Energy

Peter Williams & Saryl Jacobson

Edited by Lauri Seidlitz

Trifolium Books Inc.
Toronto, Canada

Trifolium Books Inc.
250 Merton Street, Suite 203
Toronto, Ontario, Canada M4S 1B1

© 2000 Trifolium Books Inc.

Copyright under the International Copyright Union. All rights reserved. This book is protected by copyright. Permission is hereby granted to the individual purchaser to reproduce the select pages in this book that are so specified for non-commercial individual or classroom use only. Permission is not granted for school-wide, or system-wide, reproduction of materials. No part of this publication may be transmitted, stored, or recorded in any form without the prior written permission of the publishers.

Any request for photocopying, recording, taping, or for storing on information storage and retrieval systems of any part of this book (except for the pages so designated) shall be directed in writing to the Canadian Reprography Collective, 6 Adelaide Street East, Suite 900, Toronto, Ontario, M5C 1H6; (416) 868-1620, fax (416) 868-1621.

Care has been taken to trace ownership of copyright material contained in this book. The publishers will gladly receive any information that will enable them to rectify any reference or credit line in subsequent editions.

Special Note: This resource has been reviewed for bias and stereotyping.

Canadian Cataloguing in Publication Data

Williams, Peter, 1942-

Take a technowalk to learn about mechanisms & energy

(Springboards for teaching)

ISBN 1-55244-004-4

1. Simple machines — Study and teaching (Elementary).
2. Power (Mechanics — Study and teaching (Elementary).
I. Jacobson, Saryl. II. Seidlitz, Lauri, 1966– . III. Title. IV. Series.

TJ158.W54 2000 372.3'5 C00-930343-X

531.6
.W56
2000

Project editor: Lauri Seidlitz
Design, layout, graphics: Heidy Lawrance Associates
Project coordinator: Jim Rogerson
Production coordinator: Heidy Lawrance Associates
Cover Design: Fizzz Design Inc.

Printed and bound in Canada

10 9 8 7 6 5 4 3 2 1

Trifolium's books may be purchased in bulk for educational, business, or promotional use. For information, please write: Special Sales, Trifolium Books Inc., 250 Merton Street, Suite 203, Toronto, Ontario, Canada M4S 1B1, telephone (416) 483-7211, fax (416) 483-3533, e-mail trifoliu@ican.net.

This book's text stock contains more than 50% recycled paper.

What's New?

If you would like to know about other Trifolium resources, please visit our Web Site at:

www.pubcouncil.ca/trifolium

Acknowledgments

The Publisher acknowledges with gratitude the financial support of the Government of Canada through the Book Publishing Industry Development Program (BPIDP) for all publishing activities.

The Authors would like to make special mention of the efforts of Trudy Rising, Jim Rogerson, and Lauri Seidlitz in developing this project.

Thank you all.

Safety: The activities in this book are safe when carried out in an organized, structured setting. Please ensure you provide to your students specific information about the safety routines used in your school. It is, of course, critical to assess your students' level of responsibility in determining which materials and tools to allow them to use.

Note: If you are not completely familiar with the safety requirements for the use of specialized equipment, please consult with the appropriate specialty teacher(s) before allowing use by students. As well, please make sure that your students know where all the safety equipment is, and how to use it. The publisher and authors can accept no responsibility for any damage caused or sustained by use or misuse of ideas or materials mentioned in this book.

Contents

Meet the Authors

Peter Williams

Peter Williams was educated in England, where he obtained a B.Sc. in Chemistry. Following a family tradition, he went into teaching, starting his career in a school just outside of London. Canada called, and he spent three years teaching science in Toronto schools. Australia was next on the list and he spent two years there teaching in Melbourne and the small town of Wodonga.

Peter continued his first love of science teaching upon returning to Toronto, moving from high school to elementary school and then into a consultant position. He was seconded to the Faculty of Education at York University where he developed the one-year Math, Science and Technology program for aspiring teachers. He returned to the Toronto Board of Education as the Coordinator of Science, where his responsibilities included science education from kindergarten through high school.

A published author of a highly successful science textbook series, a children's science activities book, *Light Magic,* and many teacher resource documents at the local and provincial levels, Peter continued to promote his passion for science and technology education for all students. His work in helping teachers has been recognized provincially with the "Jack Bell Award for Science Educators" and the "Emeritus Award," both given by the Science Teachers of Ontario.

Recently retired from the Toronto District School Board, Peter continues to work at developing materials for both teachers and students based on the new Science and Technology curriculum.

This latest edition of *Take a Technowalk* supports the curriculum and reflects Peter's strong belief in a hands-on approach to education.

Saryl Jacobson

Saryl grew up in suburban Montreal at a time when girls were not permitted to take shop courses in high school. Her first experience with technology was, as a young child, building cardboard furniture that was so intricate that even the drawers opened. In 1975, she enrolled in a B.Ed. program in English and Industrial Arts (now called Design & Technology). There were 60 men and 6 women in the shop course.

For the next fifteen years, Saryl taught Design & Technology, in combination with science and mathematics. In 1980, Saryl became a member of The Wood Studio, a large woodworking co-operative, and her career as a self-employed cabinet-maker flourished. Saryl has taught students from Grade 2 to adults. She was a Mathematics, Science, and Technology Consultant for the Toronto Board of Education before being promoted to Vice-Principal of a large urban elementary school. For the past three years, Saryl has been a Principal at a large mid-town Toronto school with students from Junior Kindergarten to Grade 8.

Saryl has written math, science, and technology curriculum for the Toronto and Metropolitan Toronto Boards of Education, for Gateway to Knowledge, and for Trifolium. Her collaboration with Peter Williams on the *Take a Technowalk* series has been particularly satisfying. "Peter and I share the pleasure of exploring technology with our students. This book has been our way of helping other teachers experience the same enjoyment and success."

Saryl's claim to fame is her sidekick Luka, a 12½ year old golden labrador/retriever, who has come to school with her every day, first as a classroom puppy and now as an administrator. Luka is an eager and willing participant on any *Technowalk!*

INTRODUCTION

What Is a TECHNOWALK?

What is a TECHNOWALK? Who can go on one? What do you do? It is exactly what it sounds like — a walk with your students through your community to explore the technology that you find there. You have probably taken your students on nature walks around the school, to look at the natural environment. You may also have taken them on longer trips, to a museum, an historical site, a zoo, or industrial plant. Afterwards, students return to the classroom bubbling over with enthusiasm, full of ideas and questions, working together, willing and eager to follow up with further projects. A TECHNOWALK, looking for types of mechanisms and energy at use in the community, can open up a whole new world of discovery and invention.

TECHNOWALKS provide a wonderful excuse to step outside, stretch your legs, and breathe some fresh air. Each time you venture outside, you will focus on something new. Actually, not really new, but new in terms of how you will be looking at it. You will find yourself developing an awareness of technology that you probably did not have before. By looking at only one part of something at a time, separating each part from the whole, you and your students will gain a new ability to analyze our technological world.

So, take a TECHNOWALK with your students. Leave the classroom for an hour, and see what you can find in your local community. There is no need to schedule a field trip months in advance, no need to wait for a bus, no need to rush through a curriculum topic or to try to rekindle interest in something that was completed weeks ago. TECHNOWALKS can be spontaneous events, an extension of a lesson in language or history or mathematics or science, or a way of leading into topics in any of those same or other subjects. Whatever ways you decide to use TECHNOWALKS within your curriculum, we are certain that you and your students will enjoy the investigating, designing, testing, trouble-shooting, and problem-solving that the technological process provides.

In this book, we have provided ideas for TECHNOWALKS that we have used with our students ranging in age from six to fourteen years old.

What You'll Find in this Book

Within each Section

Each section starts with background information to make you aware of the concepts being covered, the recommended approaches to use, and evaluation strategies.

The Mechanisms and Energy sections each contain plans for a sequence of TECHNOWALKS. For each TECHNOWALK, we have made suggestions for Before the Walk, During the Walk and After the Walk, and have included ***Technotalk* worksheets** that you may want to reproduce for your students. We have also provided extension/home activities. (Get your community involved; they'll love it too!)

Look for helpful tips, suggestions, and directions to reproducible forms here!

The TECHNOWALKS

Each TECHNOWALK is presented with pre- and post-walk activities, each dealing with a particular aspect of technology. The walks are organized for your convenience as follows:

Before the Walk

These activities serve both to introduce your students to the technological concept being developed in the TECHNOWALK and to elicit their current understanding. You will also find suggestions on how to stimulate interest. Particular safety and planning issues are dealt with as well as the type of record-keeping recommended.

During the Walk

The walk itself — whether outside, within the school, or within the classroom — is outlined here. In each case, direction is provided to help students focus on the concept involved. If there are collections or other records to be made, there will be additional instructions here.

Technotalks are ready-to-use forms.

After the Walk

It is vital to give students the opportunity to discuss and reflect upon their findings from excursions such as TECHNOWALKS. You will find suggestions and activities under this heading to help you and your students derive the most benefit from your observations.

Extension Activities

There are additional activities and projects provided with each TECHNOWALK that you can use to extend or enrich the treatment of any particular topic. Some of these are especially suited to older students who may be already familiar with the concepts addressed and need further stimulation.

Watch for safety issues, information on advances in technology, and general teaching suggestions here!

What Is Technology?

Kids know a lot about technology. They enjoy taking things apart. They design and build as part of their play long before learning it has a name: technology. They become excited by the adventure of finding out how things work — and why familiar objects look the way they do.

It is this thrill of discovery that makes the treatment of technology in this book so special. Teachers who have used TECHNOWALKS have found their students carrying that thrill over into the classroom, making it much easier to deliver both concepts and process skills.

Science and Technology

It is worth taking a moment to think about the distinction between these two human endeavors. Many people think of technology as applied science, and it certainly can be. However, people used technology long before they understood the science behind the processes they were using. Egyptian wall paintings, for example, show metal workers at a copper smelter, in about 1450 B.C., about 2000 years before the discovery of the chemical process involved.

Basically, science aims to increase our understanding and ability to explain nature, while technology aims to design and develop devices and processes for practical purposes. For example, science explores how gravity and friction act, but technology has given us the wheel.

Traditionally, the teaching of technology has been associated with machines and workshops. Today, it includes computers and information technology. Another trend is the movement of technology education from the workshop into other subject classrooms.

This has been a movement that has enhanced the teaching of other subjects. The problem-solving and discovery inherent in design and technology are appropriate across the curriculum. And, since wherever there are human beings, there is technology, learning more about the world means learning more about technology. That it's fun too is a bonus!

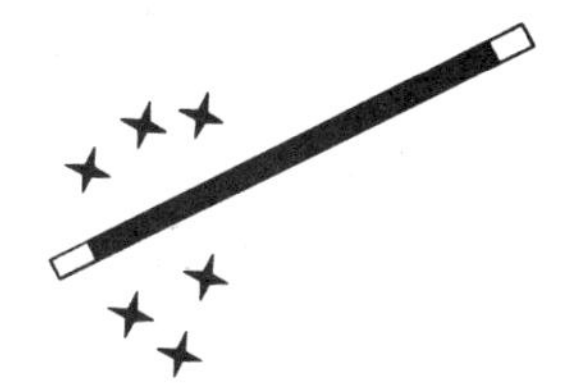

You will find a form to help you elicit your students' ideas about technology on page 76 of the Appendix. A sample reaction:

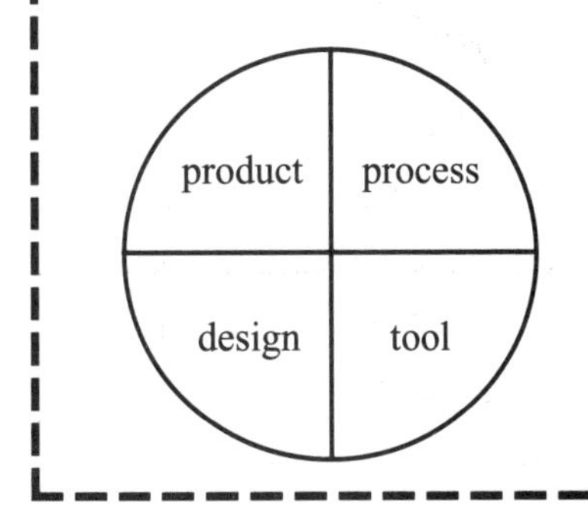

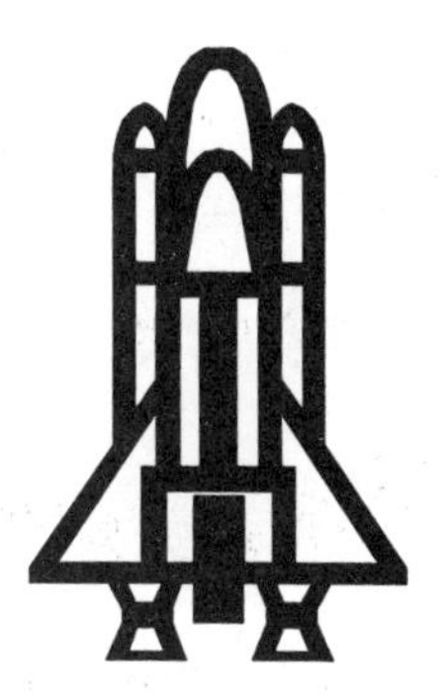

Technology can be as simple as a pair of scissors, as innovative as skates for pavement, or as sophisticated as a shuttle capable of operating in space and gliding through the air.

Making Sense of Technology

The parts that make up a technological device — materials, structures, machines, power and energy, and systems/control — are quite often right at hand and easily taken for granted. Yet most of us have asked, at some time or another, the questions on this page.

The forth and fifth questions are the focus of this book. However, all of these questions are part of the consideration of technology that you and your students will undergo during your TECHNOWALKS.

What is it made of?	material
How is it built?	structure/fabrication
What is it used for?	function
How does it work?	mechanism
What makes it go?	energy and power
What regulates it?	systems/control
How does it look?	aesthetics
How efficient is it?	ergonomics

Mechanisms

TECHNOWALKS 1-5 introduce students to the concept of a mechanism, which is something that uses or creates motion and consists of one or more simple machines. The TECHNOWALKS investigate five simple machines:

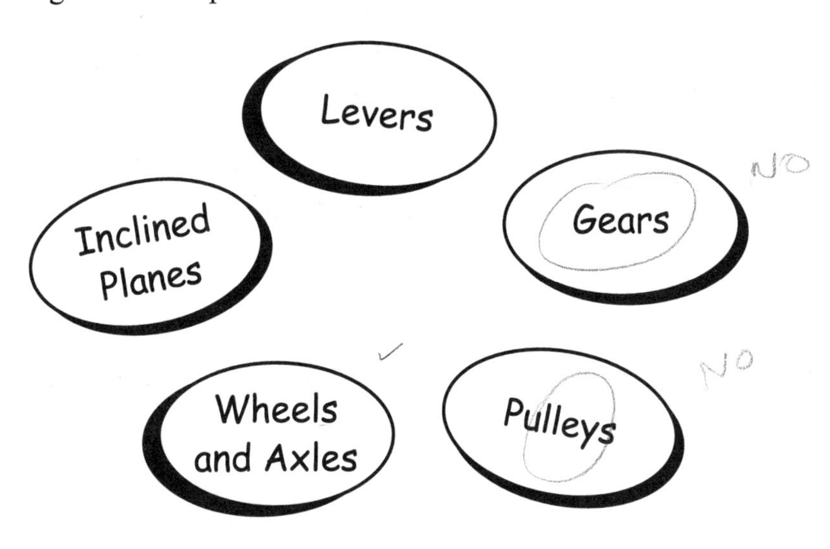

Keep an eye open for complex machines on all your TECHNOWALKS and help students identify simple machine components. By the end of the Mechanisms unit, students should be able to identify all five types of simple machines.

Energy

TECHNOWALKS 6 to 10 explore different forms of power and energy, using examples that can be found around the classroom and the school. Energy is the ability to do work and power is the rate at which energy is used. Emphasize that power and energy result in action.

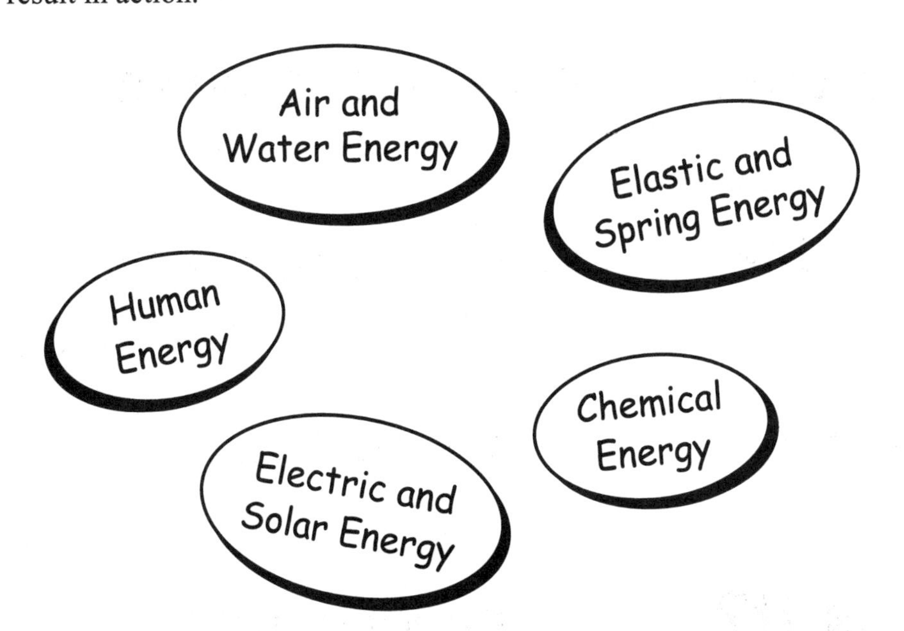

Help students develop critical thinking skills by having them assess each form of energy for is benefits, drawbacks, and uses.

Outcomes: What Do We Want Students to Gain?

Technology encourages students to learn through doing and, in the process, to become self-directed learners. Technology encourages both independent and interdependent learning. Collaborative work engages students in discussion, sharing and consideration of others. So much of what emerges is the result of people working together to plan, design, and create a product. It is often not in the success or failure of the actual product where real learning occurs, but rather in the process of working together to come to a common end result.

The following lists the general outcomes you can expect by introducing technology to your students with TECHNOWALKS.

Students will

- understand how the form, shape, color, texture, strength, and structure of a thing relates to its function and purpose.
- create and use models and pictures.
- understand the need for orderly procedures in group work and be able to develop such procedures in co-operation with others.
- be able to find more than one solution to a problem, and respect other people's solutions.
- be able to safely use simple tools and materials and to build simple objects and models.
- demonstrate an understanding of the ways in which energy is used in daily life: at home, school, and in the community.
- understand different forms of energy and reasons for using different types.
- learn about simple machines and how people use them.
- be able to explain the connections between the way people live, technology, and the environment.
- be able to describe the features of their neighborhood and the issues that affect it.

Problem-solving

An education in technology will provide students with opportunities to explore open-ended problem solving. They will learn to recognize the design procedure as a sequence for developing and completing projects in a self-directed manner. They will learn to recognize the design procedure as a problem-solving method.

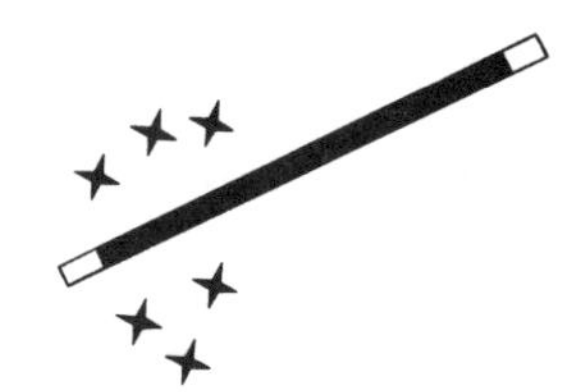

You will find detailed outcomes for each TECHNOWALK in the Appendix, pages 72-3.

Encourage students to persevere. It is always easier to discard a design that is not going well, but to stick with it and revise it until it works most often results in something far superior than first expected. Help your students recognize that making changes does not mean failure or poor designs; change will likely result in improvements in the overall design.

Curriculum Delivery Using TECHNOWALKS

Ages 6 to 8

Where to start? The primary focus from ages 6 to 8 should be to help expand children's knowledge of technology. You'll find this most easily done by providing technological experiences both inside and outside the school.

Promote the growth and development of practical communication, reasoning, and interpersonal skills through group explorations. For example, students might work together to plan and build a variety of toys using simple machines. Your school might have a display case where the toys can be displayed.

Expand their knowledge of technology.

Ages 9 to 11

From ages 9 to 11, you can expand students' explorations into other subject areas and contexts: school, family, community, and the environment. Encourage the use of correct technological terminology. Your students should begin to recognize these concepts across the curriculum and beyond. This is the appropriate age to start students using a design process to make plans and to construct products.

If you are unfamiliar with the design process as used in technology, it is not mysterious. It is simply a sequence of steps to arrive at a successful conclusion.

The **first** step in a design process is to identify and define the problem. The **second** step involves a whirl of ideas as all of the possibilities are considered. The **third** step is to select the best possibility or solution from all of the choices. Once a selection has been made, the next steps are followed in sequence, from making a plan or design, carrying out the plan or making the design, testing the plan or design and, finally, making any necessary modifications to ensure success.

Link technology to other subjects and contexts.

Some of the vocabulary you can help your students develop through their explorations are listed in the Glossary, pages 74-5 of the Appendix.

To help your students with technological vocabulary, provide opportunities for new terms to be used in the classroom. For instance, as students build, ask questions and help students express what they are doing using wider vocabulary. Post the terms you want used in the classroom, and illustrate these through drawings or photographs cut out of magazines or newspapers. Some can be written up on cards with definitions to explain their meaning.

Ages 12 to 14

When working with older students, expect them to continue to use the design process for a variety of purposes. Your students will begin to acquire specific skills related to design, such as using the proper tools and machines, drawing systems (including computer-assisted drawing programs), and creating production plans. Community TECHNOWALKS will be particularly useful to help you show students the connections between the design and technological concepts learned at school and how they apply outside the school environment.

If your students are participating in technology workshops at the local high school, incorporate this experience into your TECHNOWALKS. For example, student observations of types of machines can be woven into a workshop project.

Students use the design process to arrive at successful solutions.

Using this Book

TECHNOWALKS are short, highly focussed outings that shouldn't require much advance notice. Standard field-trip forms, signed by parents/guardians at the start of each school year, are generally all that you'll need in terms of permission to take your students outside on short walks. You may need to line up, in advance, parents who could be available on short notice. Safety is an important consideration. Ensure that your outings always have the proper ratio of adults to students.

A letter to parents, outlining the nature and purpose of the TECHNOWALKS, is an important preliminary action. It should be sent home at the start of the year so that parents are aware of the focus of your program.

Preparing for TECHNOWALKS

Choosing the Right TECHNOWALK

We have provided ideas for 10 TECHNOWALKS in this book. You will notice that they begin with mechanisms (how things work), and move to energy (what makes things go). Each new TECHNOWALK builds naturally on the others, so we recommend that you proceed in order.

Can some TECHNOWALKS be left out? Concepts explored in one are sometimes touched upon again in subsequent activities. Depending on your students' experience with technology, you could pass over one or more concepts that might already be familiar to them from other programs.

When to Use TECHNOWALKS

On pages 16-17, you will find a summary of the TECHNOWALKS in this book with suggestions about when and how to use each one. While you may find in your school that some TECHNOWALKS are better suited for K-3 or 4-6 students, many teachers have told us that each of the TECHNOWALKS is very suitable for 7-8 students, especially those who have not yet been exposed to either these activities or to technology education in general.

The TECHNOWALKS in this book have been deliberately designed so minimal advance preparation is needed. You do not, in most cases, need to arrange a bus or schedule your visits to specific locations. So sharpen your powers of observation, decide which adventure you and your students shall undertake, and get TECHNOWALKING!

TECHNOWALK Components

Alert students and their parents that the walks can take place in a variety of weather conditions. Students should dress appropriately.
Make sure that all students have provided a signed permission form. You will find a form for your use on page 88 of the Appendix.

Before the Walk

This is the time to brainstorm with your students about what they know of the technology concept to be explored and what they expect to observe on the walk. This information can be recorded on a flip chart or chalkboard so that students can refer to it later to check on their learning.

This is also a time for teaching about the concept as necessary to clear up any misunderstandings or major gaps in students' understanding.

Your students can use this time to develop their own recording sheet to take with them on the walk. Students might consider other innovative, but appropriate ways of recording the walk, such as a tape recorder or camera.

Find out what your students already know about the technological concept to be covered during the walk.

Find out what your students expect to find during the walk.

During the Walk

Each TECHNOWALK involves moving and observing; however, not all are excursions outside. Some can take place inside the school or even within a classroom.

Here students will be actively involved in observing the technology concept "in action." Encourage your students to look for all aspects of the concept and to record their observations. Help them through your questions and directions. Be prepared to hone your own powers of observation!

Plan ahead. Will you need volunteers?

Question and direct students' observations.

Consider all safety issues and be prepared.

After the Walk

This is the time to consolidate the walk. Students can check their recordings against those they predicted in Before the Walk activities. Discussion can take place to come to a common understanding about the concept observed. Clarification can be given where there is some disagreement about what was seen and how it fits into the technology concept.

Extension activities are included to further consolidate students' understanding of the concept. Extensions can be in the form of research assignments, hands-on activities to model the concept studied, or open-ended problem-solving activities to challenge students to incorporate the technology concept into their solution of the problem.

We recommend that your students work in groups during all stages of the TECHNOWALK. This allows them to support each other through discussion. The group will also help students to remind each other of safety considerations on the walk.

Evaluation and Assessment

How do you know if your students have gained knowledge and skills from the incorporation of TECHNOWALKS in your program? We have provided several ideas and tools to help you find out, including:

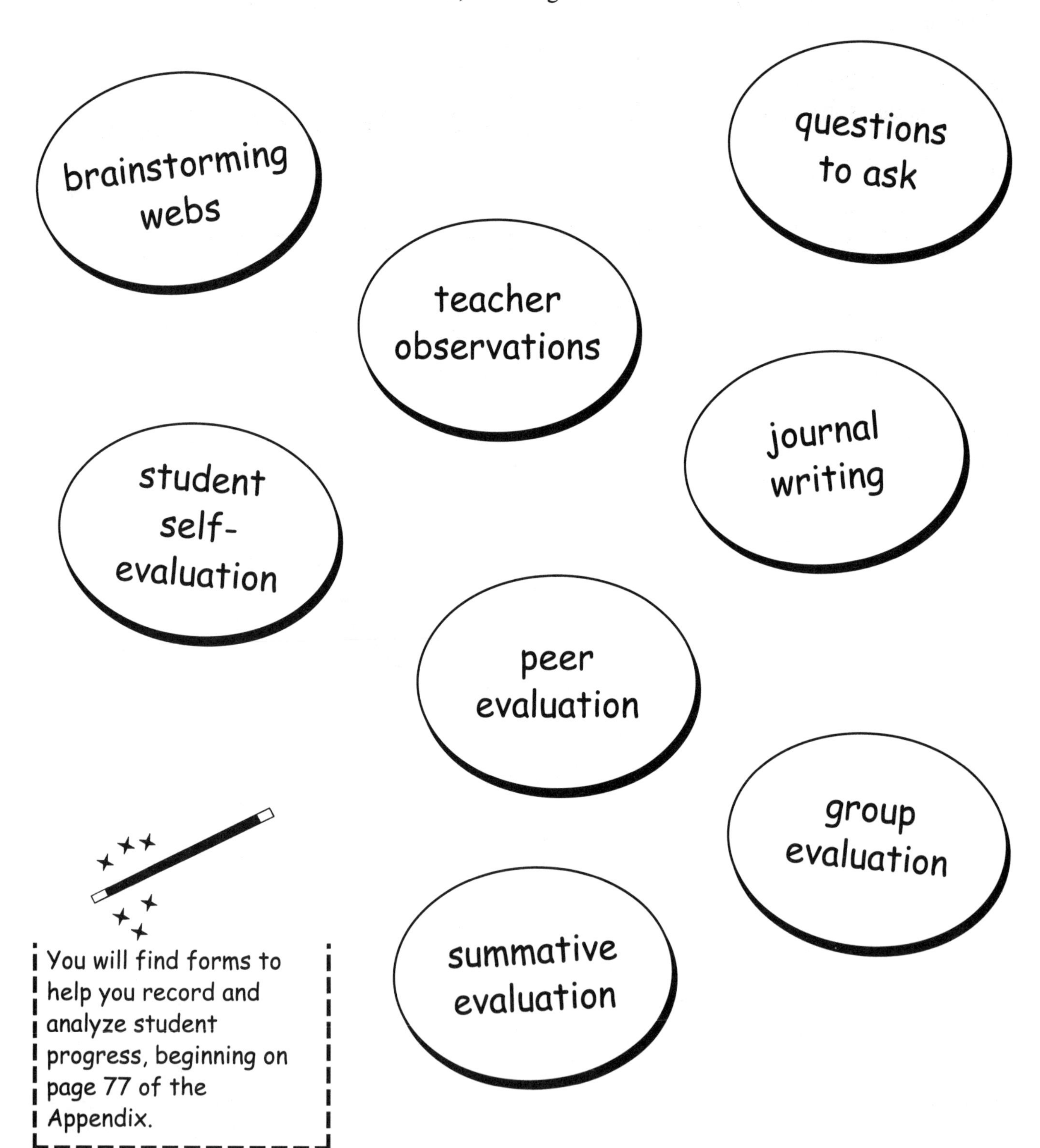

You will find forms to help you record and analyze student progress, beginning on page 77 of the Appendix.

Student Brainstorming Web

Every TECHNOWALK begins with an elicitation activity. If you wish, have students write their ideas in a web. Asking for a second web later in the unit for comparison will help you assess what students have learned. The web then becomes a summative form of evaluation. For example, the following webs show one student's growth in understanding of the concept of Mechanisms.

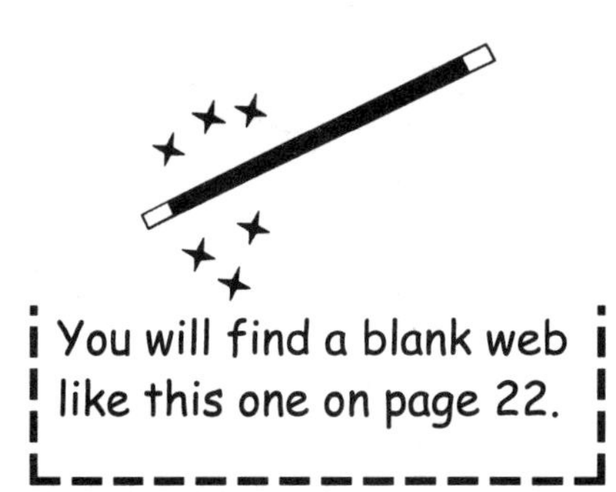

You will find a blank web like this one on page 22.

Before the Walk

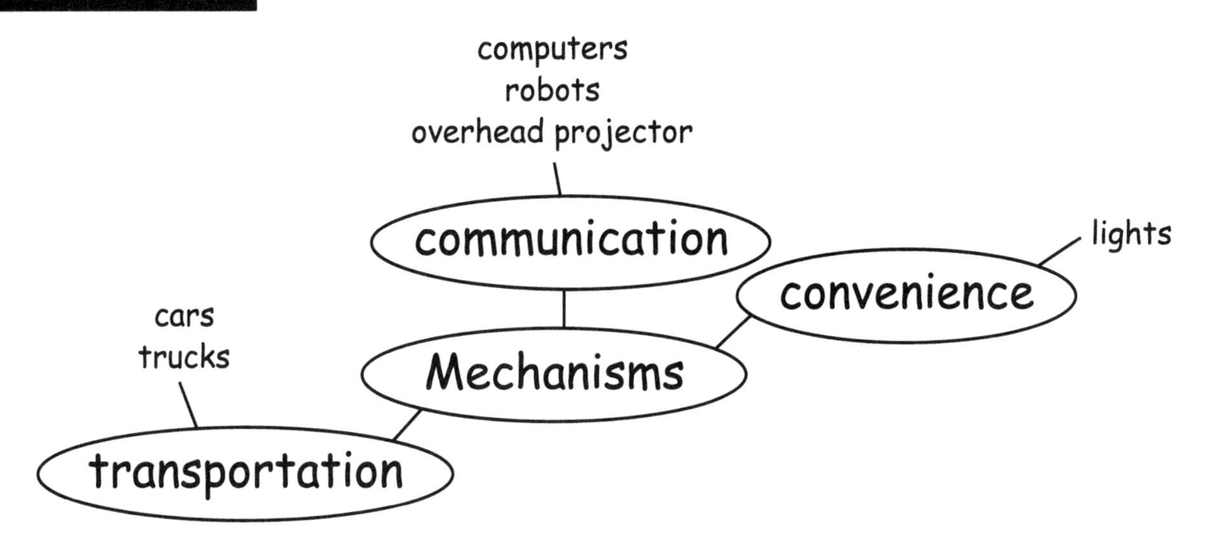

After the Walk

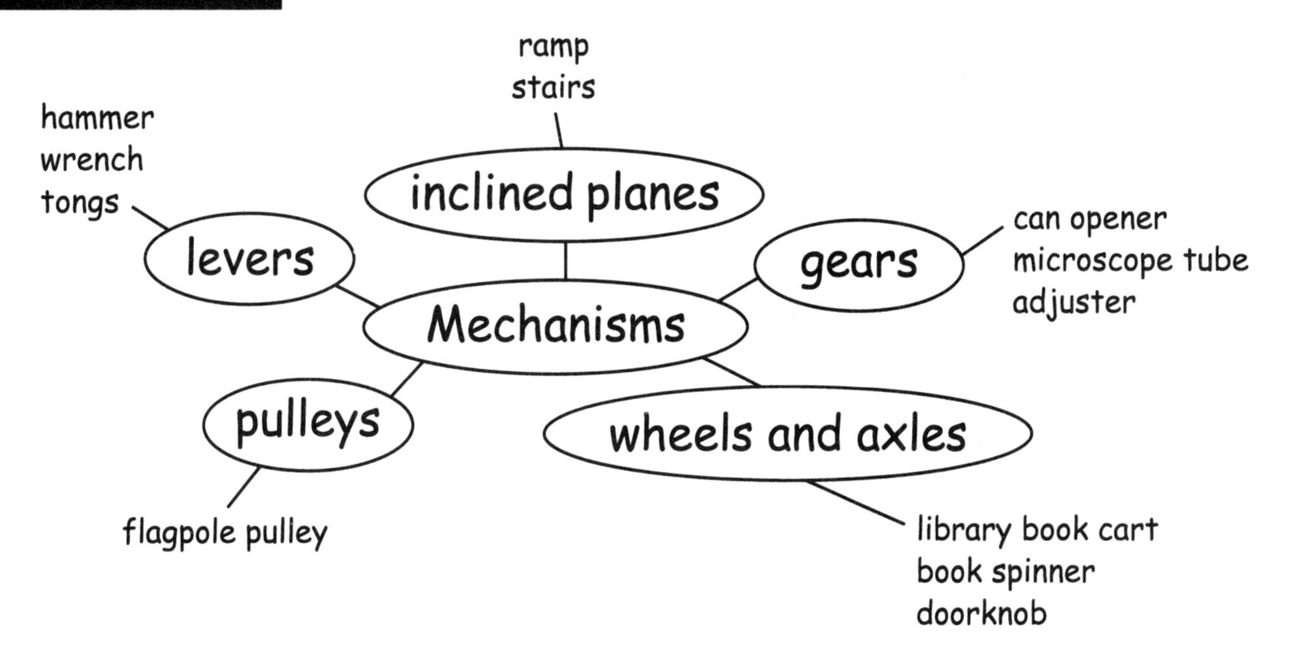

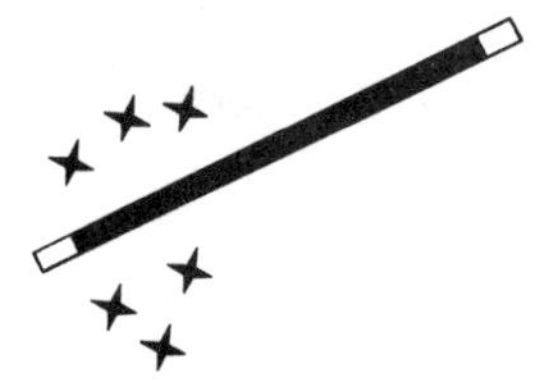

You will find forms to help you determine the success of group work and to help students assess their own performance, starting on page 79 of the Appendix.

Questions

After each TECHNOWALK, a series of questions is included to determine the level of thinking and understanding of the concepts that are developed during the TECHNOWALK.

Observations

By listening to your students as they talk to each other and to you, by looking at their recordings, and by observing them during the activities, you can assess their growth and work in technological concepts.

We recommend that you keep ongoing records of your observations together with samples of each student's work. In this way, you can identify a student's ability to work with others, interests, and areas of weakness, progress, and success.

Journals and Portfolios

Have students write in a **journal** or **log** about their accomplishments through the TECHNOWALKS. They can use this for planning, recording, and reflecting on what they have done. Teachers can also record comments here.

Student self-evaluation is an important part of the evaluation process. Encourage students to develop a **portfolio/folder** in which they can record the activities they have done, and save samples of their work. This will encourage students' articulation of their feelings, attitudes, and values with respect to technology learning.

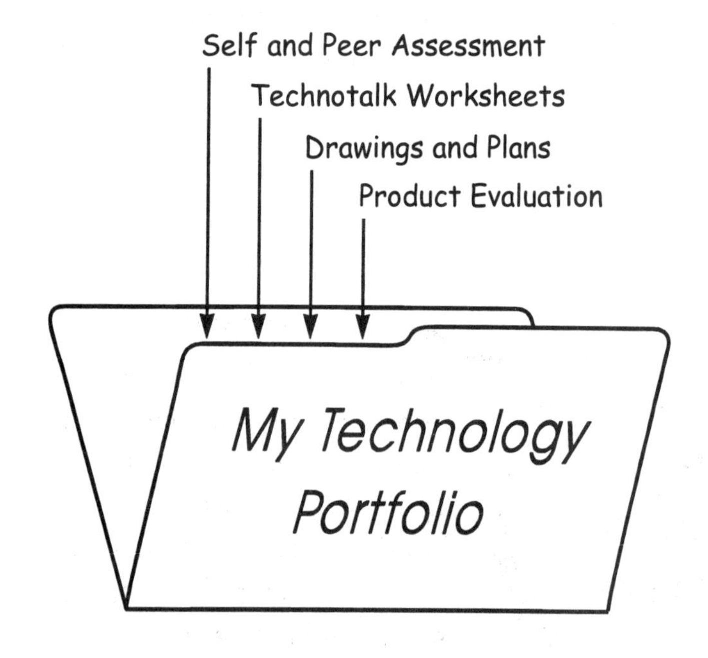

Teamwork

Because the TECHNOWALKS involve students in groups or pairs, you may wish to use the group self-assessment form supplied in this book to help you determine how effectively students felt they worked together.

Summative Evaluation

Summative evaluation, based on Bloom's Taxonomy, can be used at the end of each section to evaluate students' knowledge of technology, their understanding of the technological concepts, and their skills in problem-solving. An overview that you can use to help devise evaluation questions is provided on the next page.

Overview of Summative Evaluation

About Technology

comprehension

explain, give examples, summarize, rewrite, paraphrase, convert, distinguish, predict.

e.g., Explain, with examples, how an inclined plane can do work.

knowledge

describe, label, identify, name, state, locate, list, define, outline.

e.g., Describe three kinds of levers.

Guided Technology

analysis

analyze, break down, differentiate, discriminate, illustrate, identify, outline, point out, select, separate, sub-divide, categorize, classify, distinguish.

e.g., Make a diagram of a pencil sharpener and point out at least three wheels and axles.

application

infer, change, discover, operate, predict, relate, show, solve, use, design, build, manipulate, modify, demonstrate, compute.

e.g., Design and build a toy for a child that has a lever somewhere in it.

Problem-solving & Inquiry in Technology

evaluation

judge, compare, contrast, evaluate, criticize, justify, draw conclusions.

e.g., You are going to power a lifting machine with electric energy. Justify your choice of this form of power.

synthesis

devise, compile, design, compose, explain, organize, rearrange, plan, combine, categorize, show relationship, synthesize.

e.g., Write a short story about an imaginary explanation for how the gear was invented.

Overview of the TECHNOWALKS

TECHNOWALK	Summary
LEARNING ABOUT MECHANISMS	
1: Inclined Planes	Inclined planes allow loads to be lifted with less effort than it would take to lift them straight up.
2: Levers	Levers reduce the effort needed to move heavy loads. All levers have a load, effort and fulcrum.
3: Wheels and Axles	Wheels are similar to a continuous lever except they have axles rather than fulcrums. Wheels and axles are used to make movement of loads easier.
4: Gears	Gears are wheels with teeth that interlock. The motion of one gear cause motion in the other gears.
5: Pulleys	Pulleys change the direction of motion. Pulleys allow loads to be lifted and moved.
LEARNING ABOUT ENERGY	
6: Human Energy	Human energy can be used to lift, pull, and push. Humans use energy in different ways than they did in the past.
7: Air and Water Energy	Moving air and water can be used as sources of energy.
8: Elastic and Spring Energy	Elastics and springs can store energy to make machines work. They convert potential energy to kinetic energy. They are most practical for small items.
9: Electric and Solar Energy	Electric energy is a common and versatile form of energy. The Sun is the source of almost all energy forms.
10: Chemical Energy	Chemical energy is released during a chemical reaction. The energy is converted into heat, light, sound and motion.

Preparation	Where to Walk
books, board, toy truck or car, weights, elastic bands, ruler	neighbourhood, nearby commercial/business district
paint can, hammer, screwdriver, can opener, saw	neighbourhood, playground
thick cardboard, scissors, bulldog clips, yogurt containers, wooden shish-kabob skewers, straws, modelling clay, ruler, doorknob with one knob removed, tool box with tools including wrench	neighbourhood, park, bicycle paths, shopping mall, supermarket
bicycles	historic theme park, lumberyard, hardware store, Design and Technology shop, Family Studies room
broom handles (or hockey sticks), 3 m strong rope, commercial pulleys, string, masses	grocery store, neighbourhood, construction site, school grounds, Design and Technology shop, fitness centre

buckets, disposable cups	playground, fitness centre, gymnasium
string, pencils with erasers on the end, plastic grocery bags, syringes, rubber tubing, balloons, straws, cardboard, tape, scissors, modelling clay, paper, plastic container lids, plastic bottles	historic theme park, neighbourhood, school grounds, construction site
coffee cans and lids, masses, string, elastics, can opener, scissors, plastic bottles, masking tape, ice cream container lids, chopsticks	neighbourhood, sporting good store, classroom, office, caretaker area
drawing supplies	neighbourhood, school
matches (for use by teachers only)	local secondary school, gas station, neighbourhood intersection, school boiler room

Literature and Technology

An important role of the study of technology is to make students aware of technology as the basis of many everyday things they encounter. Literature can help them make this connection in a familiar and comfortable manner.

Through books, your students can begin to explore the many links between literature and technology. One example is the story *Mike Mulligan and His Steam Shovel,* which provides an excellent look at the way a complex piece of machinery works.

Literature is an integral part of the TECHNOWALKS in this book, especially when used before or after activities. Use literature as a basis for research as well as for an enjoyable "hook."

Consider having your students write their own stories that incorporate elements from a TECHNOWALK. For example, a story could be developed around a new invention and students' creative ideas about how to power it. Encourage students to imbed what they have learned into their creative train of thought.

You will find a list of books on pages 83-6 of the Appendix. You might wish to work together with your school librarian to discuss what the class is working on, to share your book list, and to ask for further suggestions.

Making Observations

Encourage students to use more than their sense of sight while on their TECHNOWALKS. Students should make observations using all their senses and make reasonable inferences from their observations. For example, students might hear a plane overhead or smell a truck's exhaust to note an observation of wheels and axles at work. If students see a farm truck full of cattle or horses, ask students to suggest how the animals may have been loaded into the truck (an inclined plane). If they spot a handicapped sticker on a building, ask students how the building might be accessible to people in wheelchairs (elevators that use pulleys, ramps instead of stairs). Be sure students understand the difference between an observation and an inference.

What do you see?

What do you hear?

What do you smell?

The Technology Cupboard

Even if your school has a Technology Facility or other area set aside for the tools and materials your students will need, we recommend that you maintain a Technology Cupboard in your classroom. This will allow you and your students to easily and quickly do hands-on activities, both as part of the TECHNOWALKS and as part of your other programs. Use the Technology Cupboard frequently before and after a TECHNOWALK to illustrate the concepts covered.

Some ideas of what to include are listed on the letter to parents on the next page. Check your supplies regularly and send new requests for material as needed.

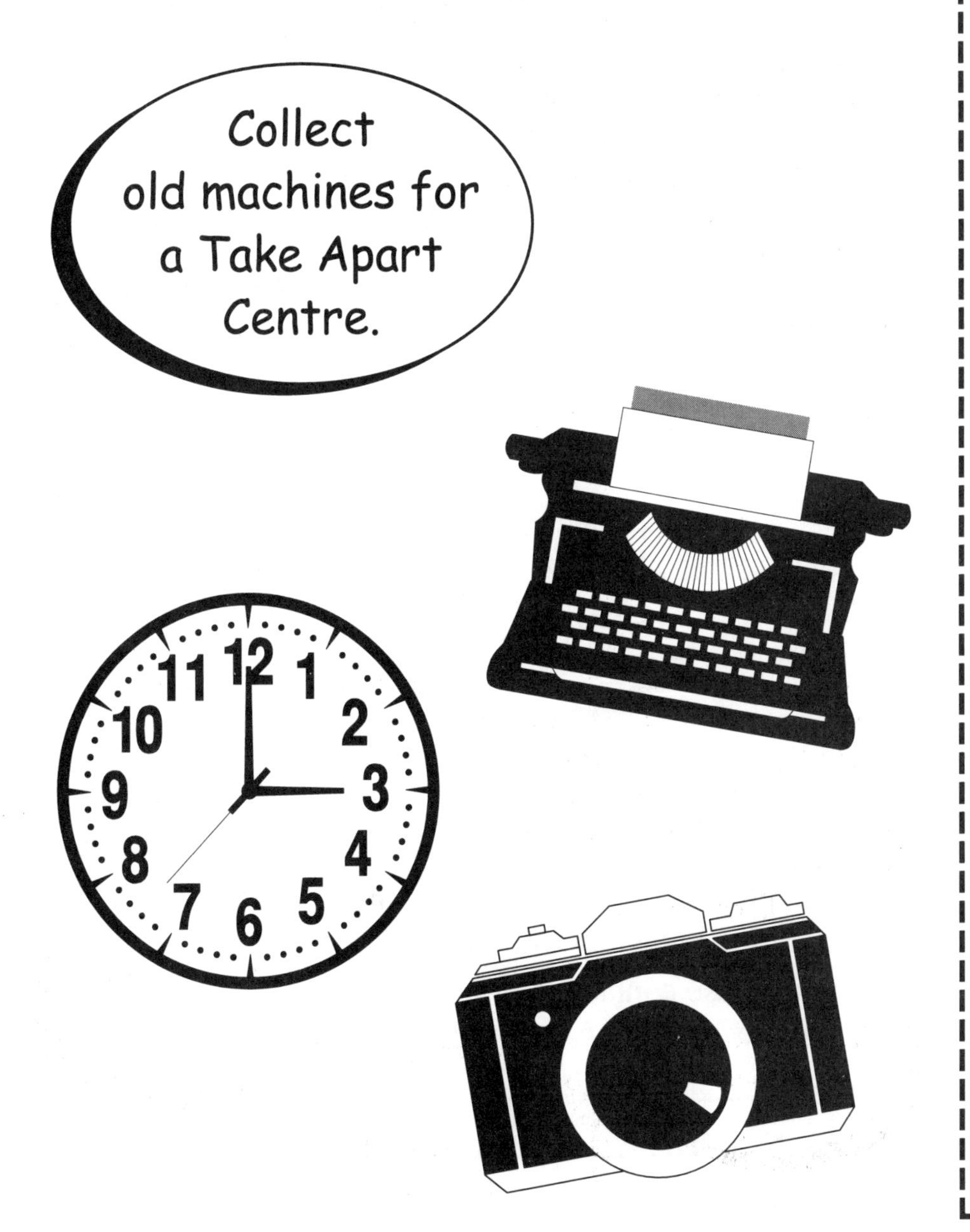

For Safety:

- Make sure cords are removed from old appliances that students may be taking apart.
- Avoid toy tools. Sharp, well-made tools are safest.
- Discard pieces of scrap wood that contain nails, glue, or finishes.
- Have students hand scissors, knives, and chisels to others with the handle out.
- Students should wear safety goggles whenever using tools.
- Work areas should be kept clean.
- Do not leave tools lying close to the edge of the bench.
- Have a safe, locked place to store tools when not in use.

Technotalk

For our Technology Cupboard, we are collecting:

- wooden planks
- pieces of flat, heavy cardboard
- boxes, blocks or bricks
- wheeled toys
- nails
- wire
- broom handles or old hockey sticks
- rope
- empty plastic bottles
- ice cream pail lids
- yogurt container lids
- Allen keys
- socket set
- empty containers with lids to store machine parts
- clamps
- corkscrews
- chisels
- screwdrivers
- pliers
- hammers
- wrenches
- saws
- paint cans (or cans with similar lids)
- garden trowels
- screws

For our Take Apart Centre, we are collecting old machines such as:

- clocks
- watches
- cameras
- bathroom scales
- record players
- egg beaters
- can openers
- garlic presses
- nutcrackers
- mixers
- lawn sprinklers
- typewriters

Dear Parents and Friends

Our class is getting ready to start an exploration of the world around us. We will be taking TECHNOWALKS to learn more about the mechanisms and energy that are part of our lives. After each walk, we will return to the classroom for hands-on activities in technology.

For our Technology Cupboard, we are collecting found materials, simple tools, and miscellaneous supplies. As well, we are starting a Take Apart Centre and are collecting old machines to discover what is inside and how they work.

Please check:

❍ Yes, please contact me about joining the class on a walk.

❍ Yes, please contact me about coming to a class to share my knowledge and experience.

❍ Yes, please contact me about donating materials for your Technology Cupboard or Take Apart Centre.

Name: ______________________________

Phone Number: ____________________

Preferred time to be contacted: Thank you!

____

Permission granted to reproduce this page for purchaser's class use only.
Copyright © 2000 Trifolium Books Inc.

Learning about Mechanisms

When you flick a pea from the end of a spoon, you are using a lever. Your finger applied the effort that made the bowl of the spoon move quickly, launching the pea into the air. A catapult works the same way. You might begin the class by reminding students of ingenious contraptions used by popular cartoon characters to capture one another. The mechanisms employed often use pulleys, inclined planes, levers, wheels and axles, and sometimes gears.

The Concept

A mechanism is something that uses or creates motion and consists of one or more simple machines that perform a specific function. We use mechanisms to make our work easier. In this section, we will explore this exciting concept in technology, find out how mechanisms work and how to design and build them. Students will learn about the five simple machines that combine to form even complex mechanisms: inclined planes, levers, wheels and axles, gears, and pulleys. Some mechanisms are so simple that we use them without realizing we are using a machine. For example, using a knife blade to open a can of peanut butter is using a lever. Using a hand-held can opener employs levers, wheels and axles, and gears. Each TECHNOWALK explores one of the five types of simple machines. On the walks, a Polaroid camera would be a great way for students to record their observations. An alternative method of recording is for students to draw what they observe. Younger students can use the *Technotalk* worksheets to record their observations. Older students should be encouraged to design their own recording sheets, or to refine the information they record on the *Technotalk* once they return to the classroom.

Select one or more books from the list on page 83 of the Appendix to introduce Mechanisms to your class. Even older students will benefit by looking through the picture books. Creative students might pair up to prepare their own stories and illustrations of mechanism use for an end-of-unit project.

Technotalk

Name: ____________________

Classifying Mechanisms

What are mechanisms and why do people use them? Categorize your ideas and place them on a web.

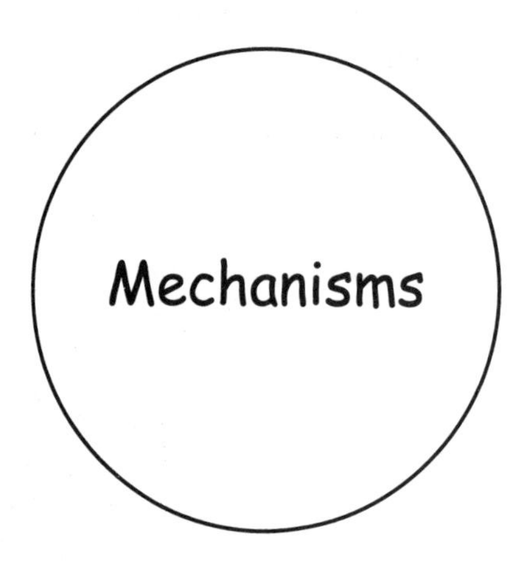

Permission granted to reproduce this page for purchaser's class use only.
Copyright © 2000 Trifolium Books Inc.

Technotalk

Name: ____________________

Taking Mechanisms Apart

Record each component of your mechanism as you take it apart.

Name of mechanism: ______________________________

Simple machine tally:

inclined planes	levers	wheels and axles	gears	pulleys

Other components:

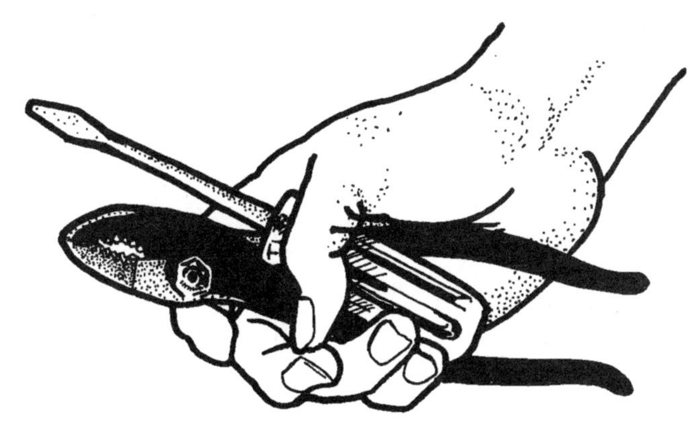

Unknown components (show sketches):

Permission granted to reproduce this page for purchaser's class use only.
Copyright © 2000 Trifolium Books Inc.

Technotalk

Name: ____________________

Home Links

At home, find as many simple machines as you can and record examples in the chart below. If you find an unusual machine, see if you can bring it to class. Look in all the rooms of your home: kitchen, bathroom, bedroom, living room, garage, work room, balcony, basement.

Simple machine	Examples at home
Inclined plane	
Lever	
Wheels and axles	
Gears	
Pulleys	

Permission granted to reproduce this page for purchaser's class use only.
Copyright © 2000 Trifolium Books Inc.

Before beginning the TECHNOWALKS, you may want to discuss some of the following principles with students:

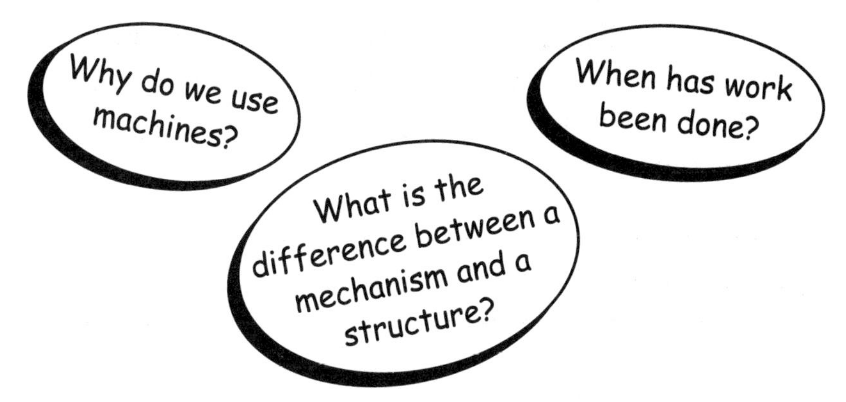

A nearby construction site can provide students with opportunities to see many simple and complex machines at work, as well as water (hydraulic) and chemical energy in action.

Work is done when a force is used to move something over a distance. Machines do not change the amount of work to be done — they just make the work easier. You might stimulate discussion by asking students which is more work in the following situations:

- lifting an elephant into a truck or walking the elephant up a plank into the truck
- climbing a flagpole to hang the Canadian flag each morning or using a pulley to raise the flag
- opening a can of paint with your fingers or using a screwdriver to pry it open
- carrying a pile of dirt with your hands from one side of the school yard to the other or using a wheelbarrow.

Be sure students understand the difference between the structure of an object and its mechanical parts. Drawing attention to the difference between the mechanisms on a car or bicycle and its chassis or frame should make this clear to most students. You might also use some of the tools in your Technology Cupboard to reinforce this concept.

Since the beginning of time, humans have used simple machines to make their lives easier. All of today's complicated machines are based on the same simple machines used long ago. You might ask students to try to stump you with a mechanism that you cannot find at least one simple machine component. This might be a good way to introduce the five simple machines to be studied.

Technotalk Worksheets

Classifying Mechanisms

Ask students to brainstorm all the mechanisms and machines they can think of. The *Technotalk* worksheet Classifying Mechanisms will help students make a web of their ideas. Older students can try categorize mechanisms by types of simple machine. Younger students can just brainstorm types of machines and the kinds of jobs they do.

You might suggest that students draw a picture to show how the parts of the mechanism fit together.

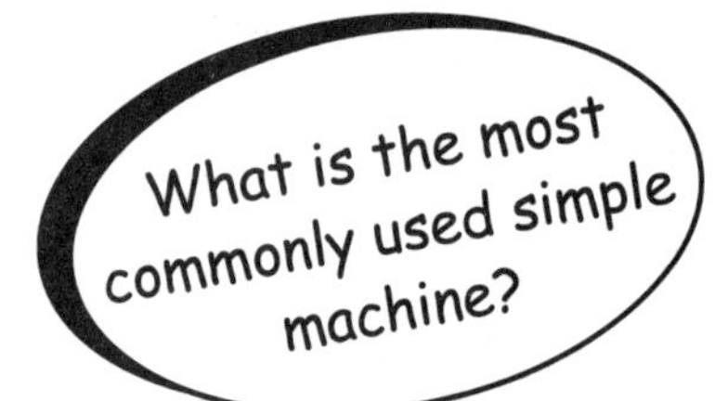

Which room of the house contains the most simple machines?

Taking Mechanisms Apart

Stimulate interest in mechanisms by allowing students to freely explore machines in the Take Apart Centre.

1. Place students in small groups and ask them to choose a mechanism to explore. Simple household utensils such as egg beaters, can openers, rotary cheese graters, and garlic presses provide good mechanisms for introductory explorations. Encourage students to examine the operation of their mechanism before they take it apart.
2. Instruct students to carefully take apart a mechanism and put the pieces on a sheet of newspaper.
3. Students should identify and count the parts that are simple machines, recording them in the chart on the *Technotalk* worksheet Taking Mechanisms Apart.
4. Instruct students to try put the pieces of their mechanism back together using their diagram.

Machines at Home

Have students use the *Technotalk* worksheet Home Links to record examples of simple machines at home. Some examples of machines students might find:

- **inclined plane** (including wedges or screws): stepladder, nail, thumb tack, door wedge, grater, bottle opener, knife, meat grinder, drill bit, corkscrew, zipper, garden hoe
- **lever:** nutcracker, hammer, chisel, wrench, tongs (barbecue, ice), spatula, tweezers, pliers, car jack, light switch, bicycle brake handle
- **wheel and axle:** door handle, stroller, toy cars, faucet, steering wheel, fan blades, dishwasher blades, washing machine pulsator, dryer drum, record player turntable
- **gear:** can opener, bottle opener, clockwork toy, watch, clock, egg beater, hand drill
- **pulley:** clothesline, drapery pull

To conclude, work with students to make a bar graph of the class results.

Extension Activity

Complex Machines

As a concluding activity for the Mechanisms unit, you might ask students to draw or cut pictures from magazines of complex machines. Have students label the simple machine components. Create a collage of complex machines for the classroom.

1 INCLINED PLANES

TECHNOWALK

Before the Walk

Stress that machines do not have to be very sophisticated. One of the simplest of the simple machines is the inclined plane. There are three kinds of inclined planes: ramp, wedge, and screw.

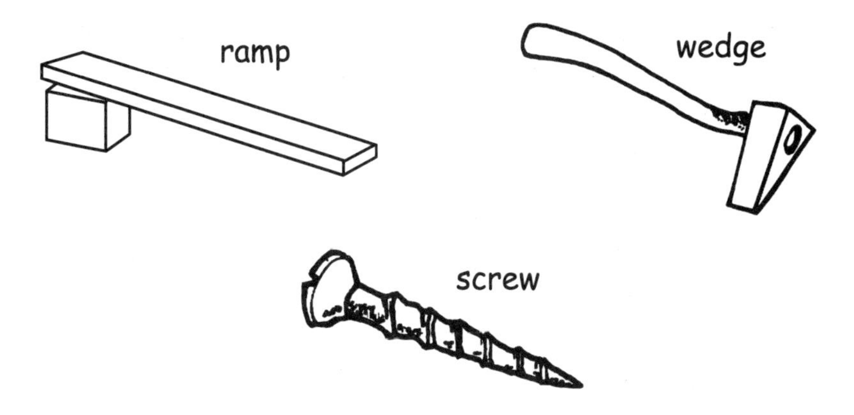

A ramp is called a machine because it allows a load to be lifted with less force or effort than it would take to move it straight up. It does this by raising the load to the desired height by moving it diagonally. The distance the load travels is greater but the amount of effort needed to move it is less.

Besides the inclined plane as a ramp, the action of an inclined plane can also seen in screws (have students imagine an inclined plane wrapped around a cylinder) and wedges, which are two inclined planes together. The inclined plane on a screw helps lift wood around the screw, making work easier as the screw moves into the wood. A wedge is a moving inclined plane. Wedges separate things by cutting, piercing, or splitting. Wedges are commonly seen in knives, saw blades and axes.

Technotalk Worksheet

Work with your class to brainstorm a list of things that are inclined planes. If you like, have students organize their ideas on the *Technotalk* worksheet Types of Inclined Planes.

stairs

driveways

doorstops

playground slides

screws

Where to Walk
Outside: neighbourhood or local commercial/ business district
Inside: school building

How Long to Walk
Outside: 1 hour
Inside: 1/2 hour

Technotalk

Name: ________________

Types of Inclined Planes

Record the types of inclined planes you see by categorizing them below.

ramps

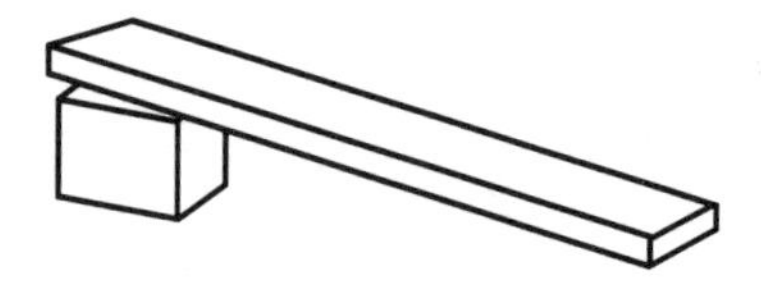

wedges

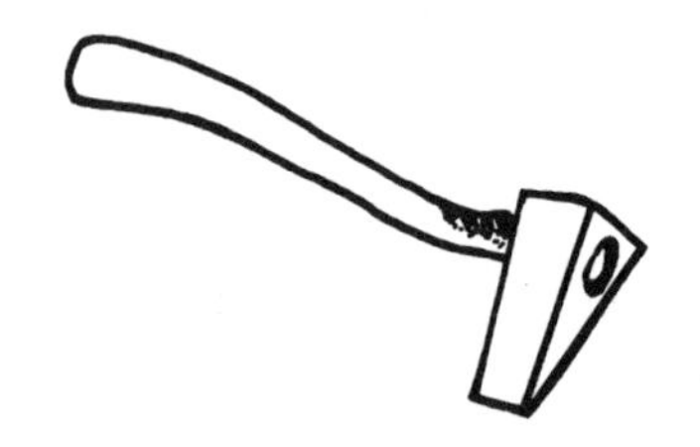

screws

Permission granted to reproduce this page for purchaser's class use only.
Copyright © 2000 Trifolium Books Inc.

Technotalk

Name: ____________________

Ramps Ahead

Find out how inclined planes make work easier with the following activity:

What You Need

- ❍ books
- ❍ board
- ❍ sturdy toy truck or car
- ❍ weights
- ❍ elastic bands
- ❍ ruler

What to Do

1. Hang the truck on the elastic band and measure the length of the elastic.
2. Place a book under the board to make a ramp. Pull the truck up the ramp with the elastic band, and again measure its length.
3. Repeat step 2 twice more, but each time add another book to make the ramp steeper.
4. Record your observations in the table.

Trial	Length of elastic band (cm)
truck hanging from band	
1 book ramp	

5. Compare the lengths of the elastic band in each trial.

Permission granted to reproduce this page for purchaser's class use only.
Copyright © 2000 Trifolium Books Inc.

Many people believe the Egyptians built the pyramids by moving huge stones up ramps of mud. Others believe they may have used levers to move stones up stairs built perpendicular to the pyramid. Workers would use levers to move a stone up slightly, reinforce this with pieces of wood, lever the other side, and so on, until the stones were gradually moved up the stairs. The Great Pyramid at Giza has 2.3 million blocks of limestone, each weighing more than 1 metric ton.

Ask students to count the inclined planes they see on their way home from school and to report back to class the next day. You might reward the student who counts (and remembers) the most.

During the Walk

1. On the TECHNOWALK, have students observe inclined planes at work. They should describe what work is being done in each case. Students can use the *Technotalk* worksheet Types of Inclined Planes to record their observations if they did not use it for the brainstorming activity.
2. Encourage students to find examples of all three kinds of inclined planes. Examples include:
   - **ramps:** stairs, escalator, ramps that make buildings accessible for people with physical disabilities, ramps for loading onto trucks, sloped driveways, playground slides
   - **wedges:** axes, shovels, saws
   - **screws:** in playground material, structures along the walk, circular staircases, top of water bottles, light bulbs.

After the Walk

1. Create a class web of inclined plane observations from the TECHNOWALK.
2. Evaluate students' understanding of the use of the inclined plane with questions such as:
   - Why is it easier to use an inclined plane to move materials rather than lifting them straight up?
   - Why do some inclined planes have rollers?

Extension Activities

Which Would You Use?

Offer younger students the following problem-solving activity. What kind of inclined plane would you use to:

- cut cake?
- secure a fire extinguisher to the wall?
- bite an apple?
- sew?
- cut paper?
- move a book without lifting it?

Ramps Ahead

The *Technotalk* worksheet Ramps Ahead provides an activity for students to explore the use of inclined planes to see how they make lifting easier. Help students analyze their results with the following questions:

- If the length of the elastic band measures the effort need to move the truck, explain how the steepness of the ramp affects the effort.
- Is there always a benefit in using a ramp?

Before the Walk

Show students a paint can, a hammer, a screwdriver, a can opener, and a saw. Ask the class how they would open the paint can. Follow their directions, which will hopefully include using the screwdriver to pry the lid off. Explain that students have just solved a technological problem using a lever as a tool.

A lever is a bar that turns or pivots at a point called the fulcrum. The fulcrum does not move. When one end of the lever is down, the other end is up and vice versa. A lever reduces the force or effort needed to move a heavy weight or load. All levers have three things in common: a load, an effort, and a fulcrum. There are three ways the load, effort and fulcrum can be arranged to make work easier:

First-class levers: Levers with a load at one end, an effort at the other and the fulcrum in between. This type of lever is used for power or precision.

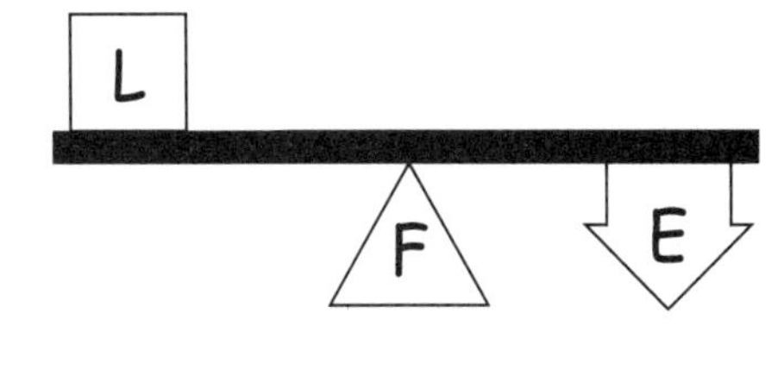

Second-class levers: Levers with the fulcrum at one end, an effort at the other end and the load in between. This lever exerts a greater force on the load than the effort.

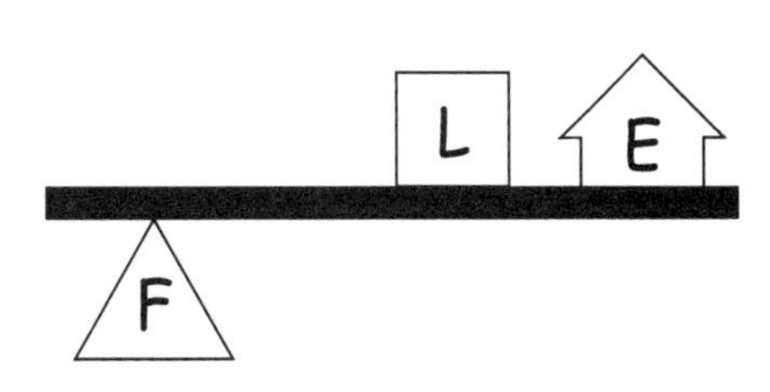

Third-class levers: Levers with the fulcrum at one end, a load at the other end and an effort in between. You must exert a greater force on the lever than the lever exerts on the load, but the load is moved quickly.

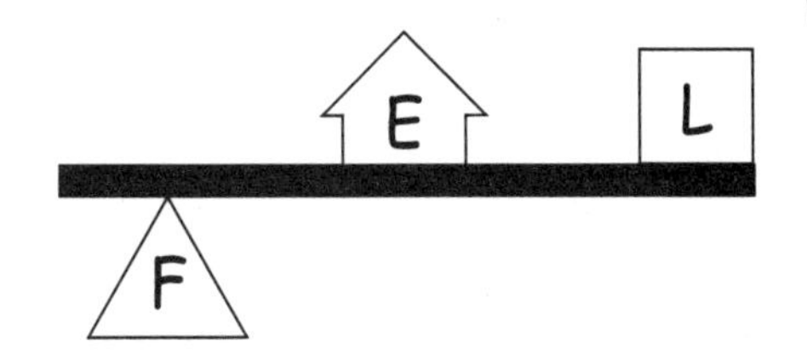

Open the paint can again, this time asking students to describe where the load, effort, and fulcrum are located on the lever you are using. Point out how far the effort-end of the lever moves compared to the load- or resistance-end of the lever. Students should see that the effort end moves much further, allowing the work to seem easier.

2

LEVERS

TECHNOWALK

teeter-totter

wheel barrow

hockey stick

nutcracker

How is your arm a lever?

What three things do levers have in common?

For what kinds of work would you use a lever?

Where to Walk
Outside: school or playground
Inside: hardware store

How Long to Walk
Outside: 1 hour
Inside: 1/2 hour

Technotalk

Name: ____________________

Types of Levers

first-class levers

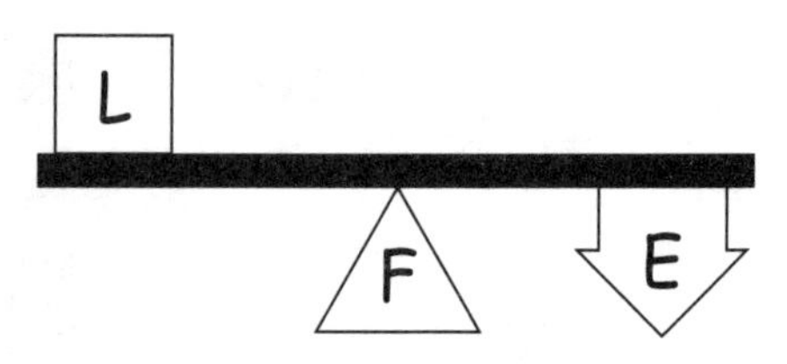

second-class levers

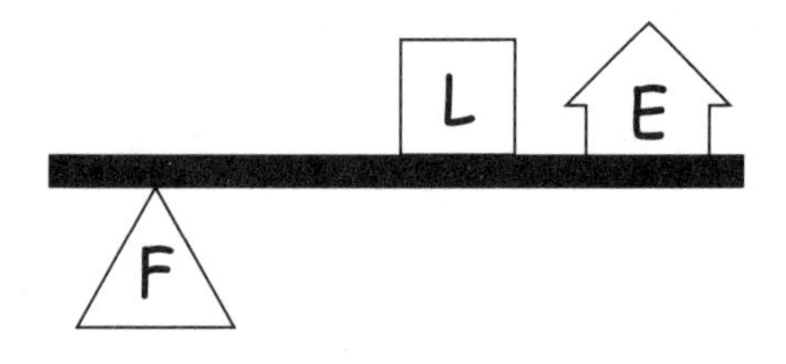

third-class levers

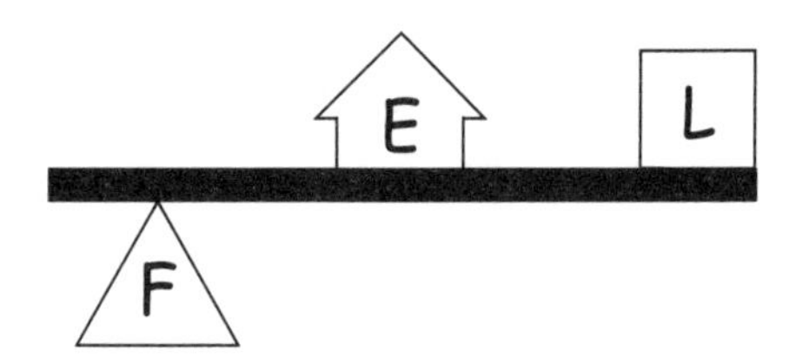

Permission granted to reproduce this page for purchaser's class use only.
Copyright © 2000 Trifolium Books Inc.

Name: ____________________

Looking for Levers at Work

Draw each object and label the effort, load, and fulcrum.

scissors

spatula

fishing rod

pliers

chopsticks

nutcracker

Permission granted to reproduce this page for purchaser's class use only.
Copyright © 2000 Trifolium Books Inc.

During the Walk

1. On the walk, have students observe and record levers at work. Encourage students to look closely at what is around them so that nothing is missed or ignored. For example, the mailbox contains simple machines. The one slot opening on a mailbox is an inclined plane to allow the letter to slide easily into the box. The larger pull-open section is both an inclined plane and a lever — the hinge is the fulcrum, the weight of the door is the load, and the handle is the effort. You can discuss why mailboxes are built this way (safety and protection from the elements). Ask student to watch for other boxes built in the same way, such as newspaper boxes or charity drop boxes for clothing.
2. Have students watch for examples of all three types of levers on the TECHNOWALK. Examples include:
   - fulcrum between effort and load (first class): teeter-totter, beam balance on a scale
   - load between effort and fulcrum (second class): wheel-barrow, bottle opener, nutcracker, standing on balls of feet, crow bar
   - effort between fulcrum and load (third class): a shovel or broom in action; sports equipment being used — a baseball bat, a tennis racquet, a hockey stick.
3. Visit a school playground to have students demonstrate different arrangements of the resistance, fulcrum, and effort to balance and load up the teeter-totter. Point out the relative distance between each component in each arrangement.

A can opener changes a small effort from a hand to a large effort that cuts through the top of a can. Strangely, the can was invented 130 years before the can opener. For many years, people used a hammer and chisel to open cans.

Why is it easier to use a lever to move materials rather than just lifting them?

What are examples of tasks you tried to do when a lever would have made a job easier?

After the Walk

Ask students to draw one example from their observations of each type of lever, indicating the effort, load, and fulcrum in each case.

Extension Activities

Looking for Levers at Work

Answers to the *Technotalk* worksheet Looking for Levers at Work:

- first class: scissors, pliers
- second class: nutcracker, spatula
- third class: chopsticks, fishing rod

3

WHEELS AND AXLES

TECHNOWALK

doorknob

merry-go-round

wagon

faucet

grocery cart

Before the Walk

The wheel is a kind of continuous lever, but instead of having a fulcrum, it has an axle. You might explain to students that a wheel and axle are really two wheels of different sizes that always rotate together. The wheel reduces the work you need to do to turn the axle because you turn the wheel over a longer distance. As the distance the wheel turns increases, the amount of effort needed decreases. Ask students to imagine a spoked bicycle wheel as being a series of levers pivoting around one fulcrum.

Allow students to examine a doorknob mechanism with one knob removed. You might challenge students' problem-solving skills by producing a toolbox and asking them to find a way to open a door without a knob. If necessary, help students select a pair of pliers and demonstrate how the pliers become a crank that replaces the knob (or wheel). A crank and axle can also be seen in a crank pencil sharpener.

How does a wheel and axle make work easier?

Compare the work of a wheel and axle to the work of an inclined plane.

Where to Walk
Outside: neighbourhood, park, bicycle paths
Inside: shopping mall or grocery store

How Long to Walk
Outside: 1 hour
Inside: 1/2 hour

Technotalk

Name: ____________________

Wheels and Axles in the Community

Record your observations on this chart.

Wheels and axles at work	Number of wheels	Observations

Permission granted to reproduce this page for purchaser's class use only.
Copyright © 2000 Trifolium Books Inc.

Technotalk

Name: ____________________

Wheels in Action

Design, build and test your own model car to investigate the action of wheels and axles.

What You Need

- ❍ thick cardboard
- ❍ scissors
- ❍ bulldog clips
- ❍ empty yogurt container
- ❍ wooden shish-kabob skewers
- ❍ straws
- ❍ modelling clay
- ❍ ruler

What to Do

1. Use the bottom of the yogurt container to draw four circles on the cardboard.
2. Cut out the circles. These are the wheels of the car.
3. Attach a bulldog clip to each end of the ruler to make the body of the car.
4. Cut a shish-kabob skewer into 5 cm lengths to make two axles.
5. Cut a straw into four 1 cm pieces.
6. Push a wheel onto the axle and add a piece of straw.
7. Put the axle through the bulldog clip add another piece of straw and wheel.
8. Repeat for the other bulldog clip.
9. If necessary, hold the wheels in place with a small piece of modelling clay.
10. Test your car by rolling it down a slope. How well did it run?
11. What changes can you make to improve the action of your car? Draw a sketch of your design ideas. If you have time, make new cars and test them.

Permission granted to reproduce this page for purchaser's class use only.
Copyright © 2000 Trifolium Books Inc.

Use the Bikes and Trikes activity to introduce the concept of friction. Friction is produced when objects rub together. Friction is a force that resists, slows, and stops motion. The rougher the surface, the more friction that is created.

Car Race

Hold a car race. Race model cars down inclined planes to find out which car is the fastest. Explain the differences in performance by comparing variables such as car size, weight, and the arrangement of the wheels.

Pop-up Card

Ask students to design a pop-up card for an upcoming special occasion. A wheel with the axle not at its centre (called a cam) can be used to turn rotary motion into an up-and-down motion.

During the Walk

On the TECHNOWALK, have students observe wheels and axles at work and play. Examples include: wheels and axles on cars, vans and trucks, steering wheels, strollers, bundle buggies, wagons, bicycles, tricycles, roller blades, skateboards, shopping carts, dollies, winch on a tow truck, wishing well, merry-go-round, street vendor carts, large garbage containers. Students will find that wheels and axles will be found on almost any system that will make it easier to move something from one place to another.

Technotalk Worksheet

Ask students to make careful observations about the wheels they see in action. Encourage students to pay attention to what the wheels are made of, what kind of surface they are travelling on, how many wheels are on the structure, how big the wheels are, and so on. Students can record their observations on the *Technotalk* worksheet Wheels and Axles in the Community.

After the Walk

Evaluate student understanding of the use of the wheel and axle by asking such questions as:

- When would you use a wheel and axle?
- Give an example showing how wheels and axles work.
- Would bigger or smaller wheels work best on racing car? Why do you think so?

Extension Activities

Bikes and Trikes

Young students can demonstrate the use of wheels and axles in the playground by riding bicycles and tricycles, or by pushing and pulling carts and strollers. They can talk about how the wheels and axles make it easier for them to get around and to move things from place to place. They might also try using wheels and axles on different types of surfaces: gravel, pavement, grass, soft dirt, hallway, and so on.

Wheels in Action

Extend the activity by asking students to test their model cars on different types of surfaces, such as carpet, concrete, tiles, wood, metal, and sandpaper. Students could then return to their TECHNOWALK observations where they saw wheels in the community to determine whether the surface facilitated movement.

Before the Walk

Gears are wheels that have teeth that interlock with each other. The motion of the first gear wheel causes a chain reaction among all the connecting gear wheels. The teeth on gears fit together like teeth on a zipper. Gears therefore transfer motion from one place to another. A gear transfers an effort over distances, making work seem easier. A gear wheel will turn a second gear wheel with the same number of teeth at the same speed, but will turn a smaller gear wheel faster and a larger gear wheel slower. These arrangements are known as spur gears. In some instances, as in the case of the bicycle, the distance between gears is so large that they are joined by a gear chain.

Other types of gears are bevel gears, worm gears and rack and pinion gears. Bevel gears are used to change the direction of motion by 90 degrees and are found in hand-held drills. Worm gears are also used to change the direction of motion 90 degrees and are found in car speedometers. Rack and pinion gears are used to change rotary motion to a to-and-fro motion.

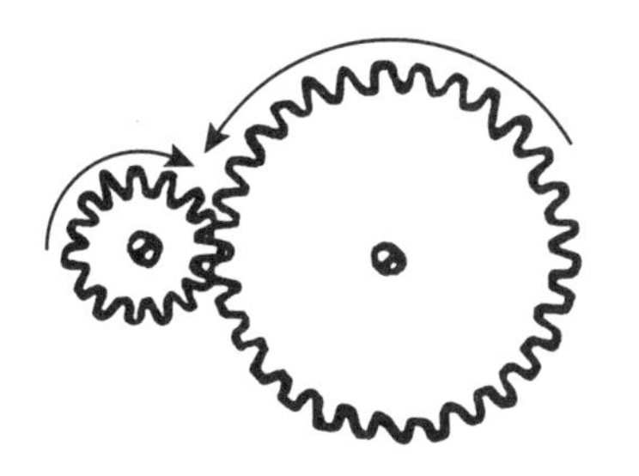

Spur gears are most often seen on bicycles.

Bevel gears are used in machines such as egg beaters.

Worm gears are commonly used in car speedometres.

Rack and pinion gears convert rotary motion to reciprocating motion. They are found in car steering mechanisms.

4

GEARS

TECHNOWALK

bicycle

can opener

watch

egg beater

wine bottle opener

Where to Walk

Outside: lumberyard, historic theme park

Inside: neighbouring school with a Design and Technology shop, hardware store, Family Studies room

How Long to Walk

Outside: 1-3 hours

Inside: 1/2 hour

Technotalk

Name: ____________________

Bicycle Gears

Bicycles have gear wheels connected by a chain. In this activity, you will investigate how the gears of a bicycle work.

What You Need

- a one-speed bicycle

What to Do

1. Turn the bicycle upside down.
2. How are the pedals and wheels connected? Draw your observations.

3. Count the number of teeth (cogs) on each gear wheel. Record your count below.

   ____

4. When the pedal goes around once, how many times does the back wheel go around?

   ____

5. Using your knowledge of how gears work, explain why the large gear wheel is on the front wheel and the small gear wheel on the back.

   ____

   ____

   ____

   ____

   ____

Permission granted to reproduce this page for purchaser's class use only.
Copyright © 2000 Trifolium Books Inc.

Technotalk

Name: ____________________

Getting into Gear

Draw arrows on both gears in each system to show which direction the gears are turning.

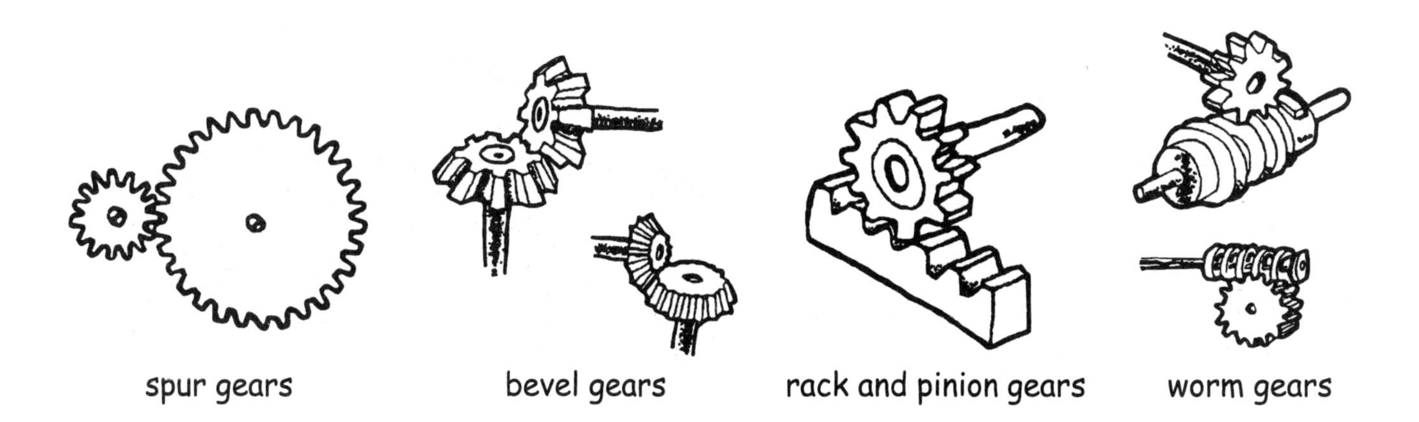

Draw your own gear system, using at least four gears. Show directions of the gears with arrows.

Permission granted to reproduce this page for purchaser's class use only.
Copyright © 2000 Trifolium Books Inc.

During the Walk

On the TECHNOWALK, you may have trouble spotting gears in action. Point out to students that gears are at work all around them, but their work is often unseen, hidden behind an object's structure. The bicycle is the most easily observed example. Other objects that have gears at work: the steering arm in a car, a car speedometer, an automatic door, a fork lift truck, an automatic garage door opener, a moving escalator. If students go to a historic theme park, there are many examples of gears at work in mechanisms such as butter churners, grinders, the tool that removes the skin from apples, and even primitive washing machines.

Gearing up means the output gear turns faster than the input gear.

Technotalk Worksheet

Stop by the bicycle rack on your walk and have students examine a one-speed bike. The *Technowalk* worksheet Bicycle Gears includes instructions for their observations.

Gearing down means the output gear turns slower than the input gear.

After the Walk

1. Discuss student observations of the bicycle as a class. Students should understand that one revolution of the large gear wheel turns the smaller gear wheel a greater number of times. This means the rider only needs to pedal once in order to get the wheels turning more.
2. At the Take Apart Centre, have students explore how gears work by looking at electric mixers, mechanical clocks and watches, and lawn sprinklers.
3. Evaluate students' understanding of the use of gears by asking:
   - Give an example to explain how gears work.
   - Describe what it means to be in high or low gear on a bicycle.
4. The *Technotalk* worksheet Getting into Gear will help you assess student understanding of the types of gears and how they work.

Extension Activities

Multi-speed Bicycles

Extend the TECHNOWALK by asking students to test different combinations of gears on a multi-speed bike. Ask students to draw all gear combinations possible for a five-speed bike. Have students choose the best possible gear for different cycling scenarios:

- riding up a steep hill
- riding as fast as possible on a flat, empty section of a bicycle path
- riding down a slight incline

Integration Opportunity

A trip to a historical theme park (for ***TECHNOWALKS*** 4 and 7) can be tied into a pioneer studies unit.

Before the Walk

Set out stations with examples of different kinds of pulleys. Allow pairs of students to visit the stations, recording their observations about how each pulley affects the kind and amount of work it does. You can purchase different kinds of pulleys at a hardware store.

Have students discuss their observations. In the discussion, be sure students understand what a pulley is and how different pulleys work. A pulley acts like a first class lever except that instead of a bar, a pulley has a rope. The pulley's axle acts like the fulcrum.

Technotalk Worksheet

The *Technotalk* worksheet Pulleys in Action provides three tasks that students need to design a pulley system to accomplish. Check student work to see that they understand how and why various types of pulleys might be used: to move an object from one place to another (moveable pulley); to lift a heavy object from one level to another (block and tackle); to move an object such as a flag or curtains (fixed pulley).

Fixed pulleys are attached to something that does not move. They change the direction of the effort force. Loads can be lifted up by pulling down on one end of the rope. Ask students why this might make work easier. (It is easier to pull down than it is to push up).

Moveable pulleys move with the load. These pulleys multiply the effort force.

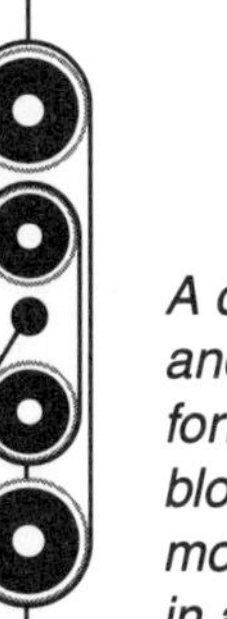

A combination of fixed and moveable pulleys form a system called a block and tackle. The more pulleys combined in a block and tackle, the less effort that is required.

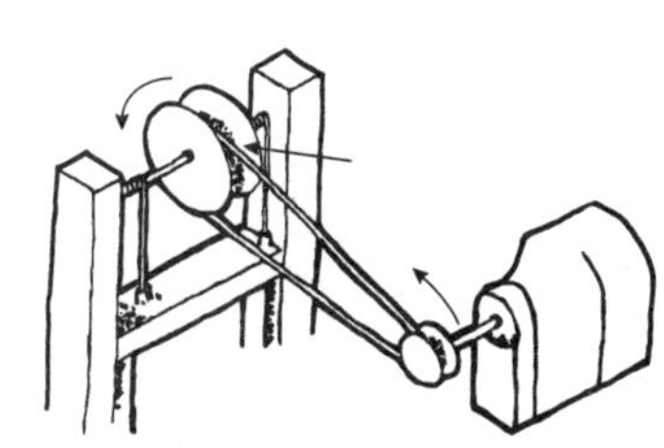

If the pulley wheels are connected by a continuous piece of rope, then they can be used to transmit power as in the case of a car engine fan belt, which is used to drive the fan, water pump and alternator.

5

PULLEYS

TECHNOWALK

single fixed

single moveable

block and tackle

continuous

A pulley is a grooved wheel that has a rope sliding over it.

Where to Walk
Outside: construction site, school grounds, residential neighbourhood
Inside: grocery store, Design and Technology shop, fitness centre

How Long to Walk
Outside: 1 hour
Inside: 1 hour

Name: ____________________

Pulleys in Action

Draw pulley systems to do the jobs listed on this page.

Move a heavy object from one place to another.

Lift a heavy object from one level to another.

Move an object such as a flag or curtains.

Draw a pulley system to perform some task inside the classroom. Be as creative as you like!

Permission granted to reproduce this page for purchaser's class use only.
Copyright © 2000 Trifolium Books Inc.

Technotalk

Name: ____________________

Pulley Tug-of-war

Make a pulley system to ensure a win in a tug-of-war game.

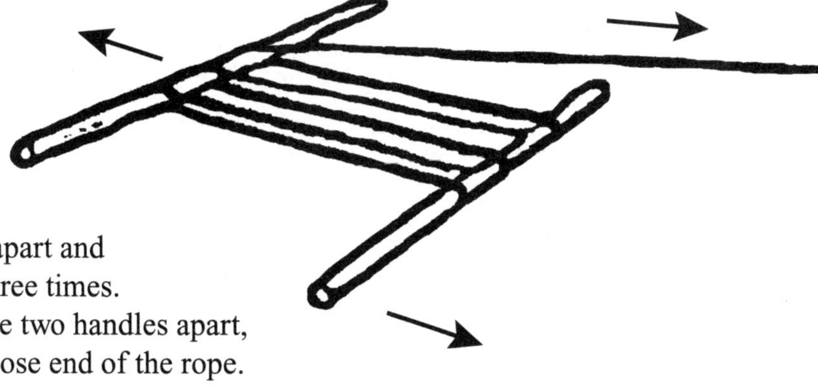

What You Need

- ❍ 3 students
- ❍ 2 broom handles (or hockey sticks)
- ❍ about 3 m of strong rope

What to Do

1. Tie one end of the rope to a broom handle.

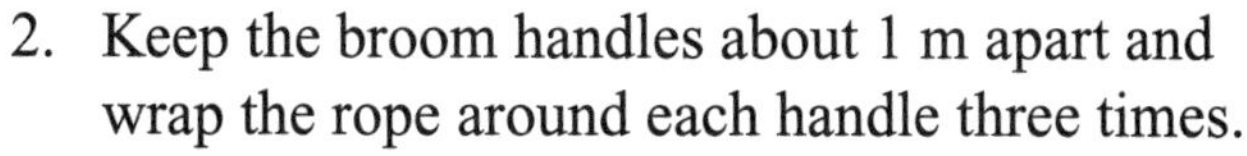

2. Keep the broom handles about 1 m apart and wrap the rope around each handle three times.
3. Choose two students to try to pull the two handles apart, while another student pulls on the loose end of the rope. What happens?

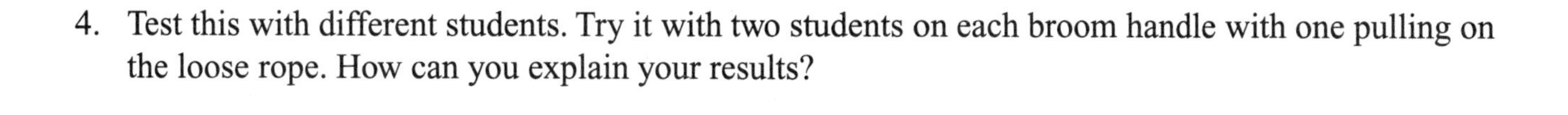

4. Test this with different students. Try it with two students on each broom handle with one pulling on the loose rope. How can you explain your results?

Permission granted to reproduce this page for purchaser's class use only.
Copyright © 2000 Trifolium Books Inc.

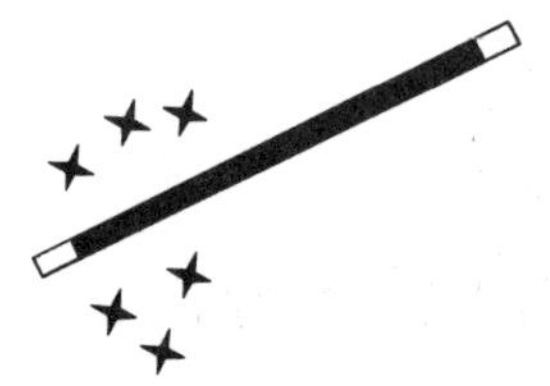

If you will be visiting a construction site, bring along pairs of binoculars to assist student observations.

Why is it easier to use pulleys to move equipment rather than just pushing or pulling them?

Give an example to explain how pulleys work.

TECHNOWALKS 5 and 6 can be linked if students visit a fitness centre. Have students observe both human energy in action as well as pulleys and other simple machines on the fitness equipment.

During the Walk

1. Begin the TECHNOWALK by visiting the school's flagpole. With the assistance of the school's custodian, have students observe the flagpole pulley and describe how it operates.
2. On the walk students should describe the kind of work done by the pulleys they observe and they should record details about the pulley system (how many pulleys are in the system, size of the pulleys, direction of effort movement and direction of result). Examples of pulleys: flagpoles, clotheslines, window washer platforms, cranes, tow trucks, sailboats, mechanic's garage (block and tackle), curtains or blinds, lifeboats on ships, gondola or ski lifts, universal gym equipment, climbing ropes in the gymnasium, conveyer belts, band saws, table saws, equipment used to raise and lower a gym's basketball backboard.
3. Have students develop an observations sheet to record the ways pulleys are being used. Suggest that students make a sketch of each different pulley system they encounter. Older students should be able to label their diagrams with the direction of the rope, the effort, and load. If students visit a fitness centre, they will be able to see many pulley systems in operation on the universal gym equipment. Ask students to observe people using the equipment, making note of other simple machines, including the body's levers, at work.

After the Walk

Ask students to describe the pulleys they observed on their walk.

Extension Activities

Pulley Tug-of-war

Use the *Technotalk* worksheet Pulley Tug-of-war to show students how much work a pulley system can do. In the first trial, no matter how hard the two students pull on the broom handles, the single student can still pull the two handles together. This works because the rope forms a simple pulley system that allows it to produce about five times more force than is being exerted by the two students pulling on the handles.

Learning about Energy

Making anything move or change, such as walking from one place to another, boiling water, or powering a lawn mower, takes energy. Energy is what makes things happen; it is the ability to cause change. Power describes how quickly work is being done in a certain length of time.

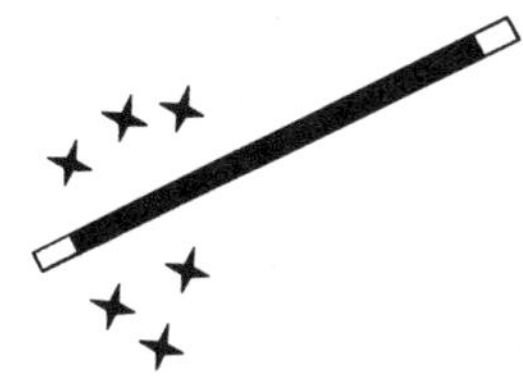

The chemical energy in plants came originally from the Sun. Because people need to eat plants or animals that eat plants, ultimately all our food comes from the Sun.

The Concept

Energy enables work to be done. Energy exists in a number of different forms: human, air and water, elastic and spring, electrical and solar, and chemical. Making things work often involves converting energy from one form to another. For example, a stove uses electrical energy converted to heat energy; a car engine uses chemical energy converted to mechanical energy; fireworks convert chemical energy to light, sound, heat, and motion. When you eat, your body changes chemical energy in food into energy that keeps your body warm and moves your muscles. When muscles do work on a machine, such as moving a bicycle pedal, the chemical energy is converted to mechanical energy.

Power describes how quickly work is being done in a certain length of time. The general understanding of energy and power is that they are interchangeable terms. People refer to electric energy and electric power as the same thing. As a formal definition, however, energy is the capacity to do work and power is the rate of doing work.

Each of the TECHNOWALKS explores one type of energy use in the community. Some are easier to spot than others. Use the abundance or scarcity of each energy source to discuss concepts such as renewable and non-renewable energy, energy conservation, human needs for technology, and costs of technology.

Look at the reading list on page 85 of the Appendix to find useful stories to introduce this unit on Energy.

Power is the rate at which energy is converted from one form to another.

Energy is the ability to do work.

Name: ____________________

Forms of Energy

Complete this web with as many forms of energy as you know. One branch has been started for you.

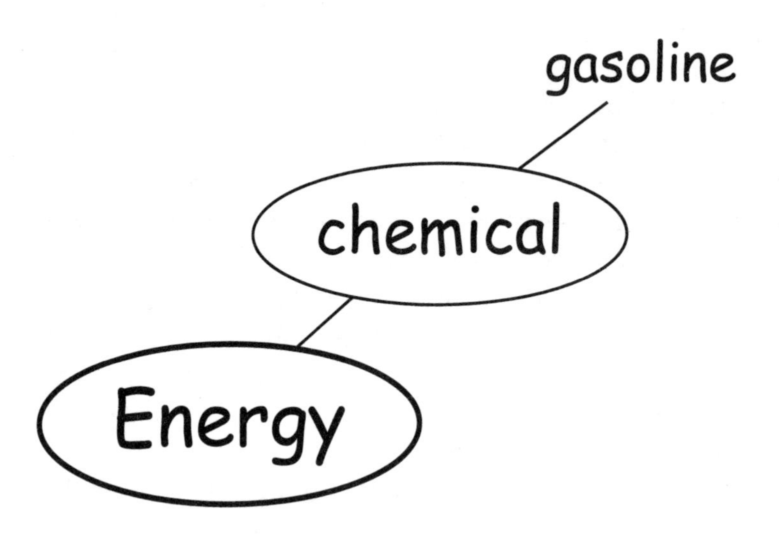

Permission granted to reproduce this page for purchaser's class use only.
Copyright © 2000 Trifolium Books Inc.

Technotalk

Name: ____________________

Energy Conversions

Complete this chart by describing the energy conversions that take place in different types of mechanisms.

Mechanism	From	To
toaster	electrical	heat
radio	electrical	sound
car	chemical	mechanical

Permission granted to reproduce this page for purchaser's class use only.
Copyright © 2000 Trifolium Books Inc.

Remind students that they have a responsibility to use all scientific equipment safely, including elastic bands. Before shooting the bands, you might ask students to stand in a single row facing a wall so that no student's band will shoot in the direction of any other student.

Things that are moving have kinetic energy. Kinetic energy can be transferred from one object to another (think of a cart rolling down a hill bumping into something at the bottom). A bowling ball transfers kinetic energy from the ball to the pins. Even if the ball only hits some of the pins, those pins gain kinetic energy that they then transfer to other pins. Dominos are another example of kinetic energy transference.

Objects can have energy when they are stationary as well. This stationary form of energy is called potential energy. (Potential energy is energy that comes from position or condition.) It is "stored energy" that can be turned into kinetic energy to cause motion.

If you need an example to illustrate these concepts, try a simple elastic band activity. Have each student hold a thick elastic band and touch it lightly to their lip. Then have students stretch the band quickly several times, again touching it to their lip lightly. The energy in the motion has been converted to heat energy that the students can feel. Now have students stretch the elastic band, which then has potential energy. Then have students shoot their bands to the ceiling or sky, releasing kinetic energy. None of this demonstration deals with power until the elastic is released, and then students can see the power (which would change if they used a heavier elastic band) over the distance it flies.

Which has more potential energy: a book on the top shelf or a book on the bottom shelf?

Technotalk Worksheet

The *Technotalk* worksheet Energy Conversions will help students understand the concept of energy conversions.

Extension Activities

Movement is the outcome of energy input.

Brainstorming

Have students brainstorm as many examples of energy as they can, categorizing them into groups. The *Technotalk* worksheet Forms of Energy can help students get started.

Swings

On one of your TECHNOWALKS, visit the school playground and have a student demonstrate kinetic and potential energy on the swing. Have students discuss when kinetic changes to potential and vice versa.

Which has more kinetic energy: a train rolling down a track or a child riding a bike down a steep hill?

Amusement Park

Have students draw a roller coaster and other amusement park rides, labelling kinetic and potential energy on each attraction.

6

HUMAN ENERGY

TECHNOWALK

Before the Walk

In the past, human energy was used extensively to do work. Even today we use human energy to do some types of work despite the development of other sources of energy. Brainstorm different forms of human energy with your students.

What kinds of machines help humans spend less energy?

What forms of energy do humans use to make machines work?

walking

carrying

bicycling

cleaning

Humans use lifting, pulling, and pushing energy to make machines move. Most machines used by humans help them spend less energy. Cars help people move quickly from place to place without walking or running; telephones allow people to communicate without having to move to be near the person they wish to speak to; the school's bell tells students to begin a new class without requiring a person to travel from room to room.

During the Walk

On the TECHNOWALK, have students observe human energy in action. Encourage students to note when humans are using machines to reduce or increase the energy they spend.

Technotalk Worksheet

Ask younger students to record their observations on the *Technotalk* worksheet Human Energy. You can help younger students organize their observations into categories after the walk. Ask older students to develop categories for the kind of human energy they expect to observe before leaving the classroom.

Where to Walk

Outside: playground where people are involved in some kind of sport

Inside: local fitness centre, gymnasium or recreation centre with people exercising

How Long to Walk

Outside: 1 hour

Inside: 1 hour

Technotalk

Name: ____________________

Human Energy

Human energy is used to …

Permission granted to reproduce this page for purchaser's class use only.
Copyright © 2000 Trifolium Books Inc.

Technotalk

Name: ____________________

Water Brigade

In teams, try to transfer water from one bucket to another without spilling a drop.

What You Need

- ❍ 4 buckets
- ❍ 1 disposable cup (Styrofoam or paper) for each player

What to Do

1. Divide into two teams.
2. The teams should stand in two parallel lines.
3. Place a bucket at either end of each team.
4. Fill the buckets at one end with water.
5. On the count of three, the first players on each team dip their cups into their buckets of water.
6. The water is passed from player to player on each team by pouring it into the next cup.
7. The last players pour their cups of water into the empty bucket.
8. The winning team is the one that finishes with the most water in the bucket
9. Try to create other ways to transfer the water using human power.
10. Write instructions for your group's ideas for step 9. Exchange your instructions with the other team and test them out.

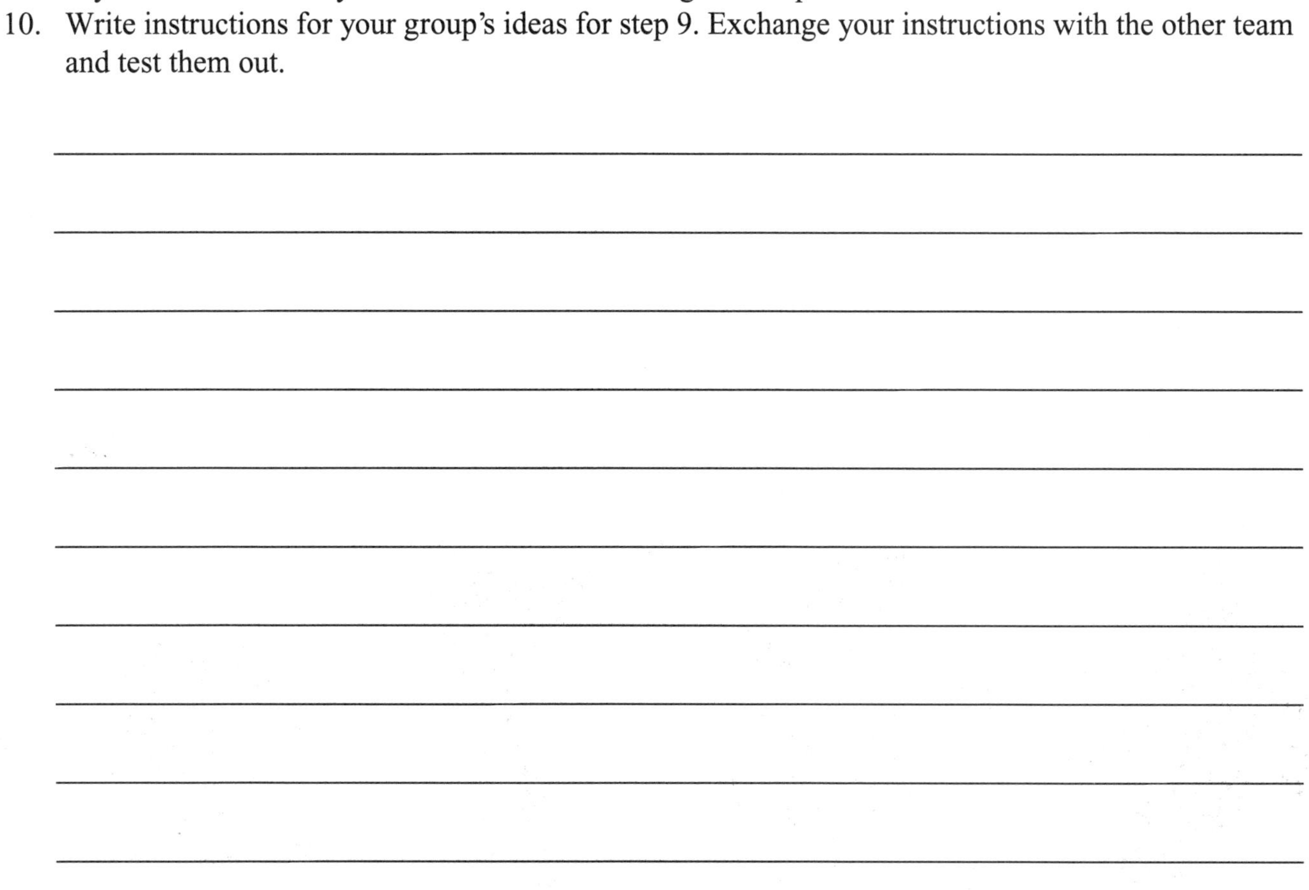

Permission granted to reproduce this page for purchaser's class use only.
Copyright © 2000 Trifolium Books Inc.

The ancient Egyptians used slave labour as a cheap source of energy to pull ploughs, till fields, dig, build, turn pumps, and carry water.

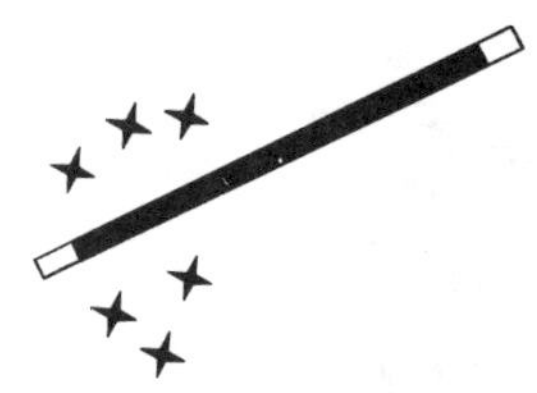

The widespread use of exercise machines is a relatively recent phenomenon in human history. You might discuss the societal changes that have occurred to make the use of such machines desirable.

Many machines help humans spend less energy. Ask students: What kinds of machines are more powerful than human energy? Which are not necessarily more powerful, but are useful nevertheless? Examples to stimulate discussion: car, typewriter, television remote control.

After the Walk

1. Collect class results from the TECHNOWALK. Help younger students organize their observations into categories. Examples include:

to get around	to move people or things	to do work
• walking	• using baby carriages	• using tools

   or

pushing something	pulling something	lifting something
• shovel snow	• sweeping leaves	• weights

   Ask older students to combine their results in groups to see if they can improve their categories.

2. Evaluate students' understanding of the use of human energy by asking:
   - When is human energy used rather than some other kind of energy?
   - What are examples of your own use of human energy in a typical day?
   - What kinds of machines save human energy?

Extension Activities

Human Power: Past, Present, Future

Explore the use of human energy in the past, present, and future with your older students. You can ask them to make comparisons and judgements about whether there have been improvements over the years in how humans use their energy. Ask students to work in small groups to list work done by humans, how the work was done in the past, and how it is done in the present. Ask students to speculate how the work might be done in the future. For example:

Human Power: Past, Present, and Future

Work	Past	Present	Future
washing clothes	scrubbing by hand	washing machine	self-cleaning clothes
washing dishes	washing by hand	dish-washer	disposable dishes

You might raise energy conservation issues as a class discussion associated with this activity. Ask students to weigh present and future uses of energy in terms of benefits to society and convenience.

Water Brigade

In the *Technotalk* activity, Water Brigade, students should find that a team effort is often needed to make the use of human energy the best approach to a challenge.

7

AIR AND WATER ENERGY

windmills

erosion

sailboats

hydraulics

TECHNOWALK

Before the Walk

Air and water are not by themselves sources of energy, but the movement of air and water can be very powerful. Air can be used as a source of energy in several ways, including windmills that convert wind energy to mechanical energy used to pump water and grind corn, and wind-powered vehicles such as sailboats, hot-air balloons, or windsurfers. Energy from the wind is also converted to electrical energy in windmills.

Water energy can be used in several ways. It can come from falling water, as in a water wheel. The water at the top of the falling stream of water has potential energy. As the water falls, the potential energy is converted into kinetic energy. The kinetic energy from a water wheel can be used to drive other machinery.

If you have a photograph of the Grand Canyon, you might explain that it was formed through the erosive effects of water motion over many thousands of years. You might ask students to consider how much human energy it would take to dig a hole the size of the Grand Canyon.

Why would a region decide to use wind or water energy rather than some other energy source?

Why would one form of energy (such as wind) be converted into another kind of energy (electricity)?

When would wind and water energy not be a good choice for a region?

In preparation for the walk, ask each student to make a "wind-tester" by tying a thread to a tack and sticking the tack into the eraser on a pencil. Alternately, you could supply each student with a plastic grocery bag that will "balloon out" in a strong wind.

Where to Walk

Outside: historic theme park, construction site, near a lake or ocean, school grounds

Inside: you might show a video of pioneer life showing air and water power being used

How Long to Walk

Outside: 1-3 hours, depending on where you walk

Technotalk

Name: ____________________

Finding Evidence of the Power of Air and Water

Fill in the chart with examples of evidence you observe and how you can infer that air or water energy was there.

Observed evidence	Inference

Permission granted to reproduce this page for purchaser's class use only.
Copyright © 2000 Trifolium Books Inc.

Name: ____________________

School Wind Test

Draw the outline of your school below and test the relative strength and direction of the wind from at least six spots. Try to test the wind near some straight walls and some corners.

N

Permission granted to reproduce this page for purchaser's class use only.
Copyright © 2000 Trifolium Books Inc.

Choose a windy day for your walk!

While it may be difficult to observe water energy in action, as in the case of water wheels in hydroelectric systems, hydraulic action is fairly common. Hydraulics is the controlled movement of water from one place to another. You can demonstrate this with two syringes connected by rubber tubing. When one syringe piston is pushed in, the other syringe piston is pushed out by the pressure of the water. Explain to students that this motion can be used to do work. In some machines, oils replace water. Observe hydraulics in the motion of rams on machines such as bulldozers, excavators, and hydraulic lifts in garages.

During the Walk

On the TECHNOWALK, ask students to look for signs of air and water energy in action. Students should consider how people use air and water energy to get around, to generate power, and to do work. Since observing water energy at work may be difficult, suggest that students look for evidence of air and water energy, such as an eroded river bank, leaves blowing across the school ground, flags flying, a collection of debris near an ocean front, a collection of leaves and twigs near a storm sewer, a tree growing in the direction of a prevailing wind, leaves and garbage collected in building corners, and so on. The *Technotalk* worksheet Finding Evidence of the Power of Air and Water can help students record their observations and inferences.

Technotalk Worksheet

Have students sketch an outline of the school on their *Technotalk* worksheet School Wind Test. Ask them to test the wind using their wind-tester at a variety of spots around the building. They should label their drawing and mark their observations of wind strength and direction on their sketch. Bring a compass to tell students the directions of north, south, east, and west. If possible, provide older students with their own compasses and tape measures and encourage them to make their drawings to scale.

After the Walk

1. Discuss as a class all the evidence students found for air and water energy. Encourage students to show reasonable thought in their inferences. Discuss reasons why wind and water energy might not be more widely used.
2. Ask students to use their drawings of wind tests around the school to suggest solutions to the following problem: If the school wanted to save money on its electric bill by installing a windmill to generate electricity, where should they locate it? Ask students whether this would be a practical option for the school. (Is the school in an area with strong, reliable winds?)

Extension Activity

Wind-powered Vehicles

Provide students with a variety of materials and challenge small groups to design and build wind-powered vehicles. You might provide: balloons, straws, cardboard, tape, scissors, modelling clay, paper, plastic container lids, plastic bottles cut in half (for boats), paper fasteners, hole punch, wooden shish-kabob skewers, and so on.

8

TECHNOWALK

ELASTIC AND SPRING ENERGY

Before the Walk

Elastics and springs can store energy that can be released to make a machine do work. The stored energy is the machine's potential energy. Elastics and springs can store energy by being twisted, stretched, or compressed. When the potential energy is released, it is converted into kinetic energy.

Generate a list with students of where they might find examples of elastic and spring energy. Students may be most familiar with elastics and springs in small toys. If possible, demonstrate a jack-in-the-box toy or joke gift with a spring-loaded "surprise" inside. Allow students to explore mechanisms in the Take Apart Centre that contain springs or elastic.

elastic bands

wound-up springs

ballpoint pens that click on

spring-loaded toys

What kinds of technological problems can elastic or spring energy not solve?

What kinds of technological problems can elastic or spring energy solve?

What are limitations of elastic or spring energy?

Where to Walk
Outside: neighbourhood
Inside: classroom, office, caretaker area, sporting goods store

How Long to Walk
Outside: 1/2 hour
Inside: 1/2 to 1 hour

Technotalk

Name: ___________________

Tin Can Toy

Use some simple materials to create a tin can toy that uses elastic energy.

What You Need

- ❍ 1 coffee can
- ❍ 2 coffee can lids to fit
- ❍ elastic band
- ❍ weight or mass
- ❍ scissors

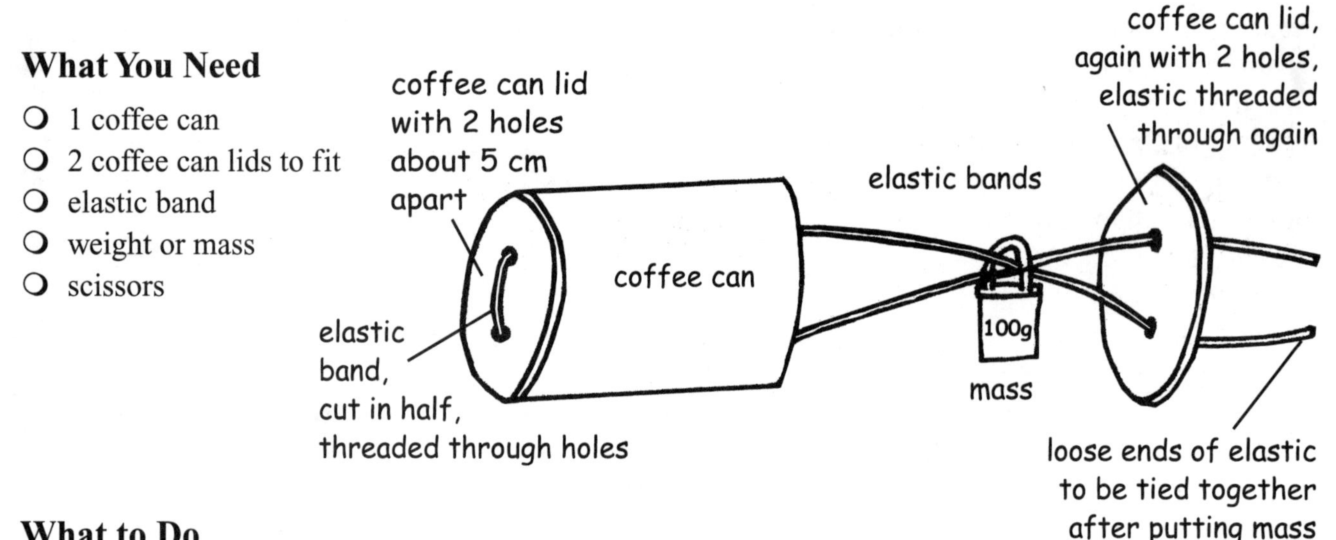

What to Do

1. Remove both ends of the coffee can.
2. With the scissors, make two holes about 5 cm apart in each plastic lid.
3. Cut the elastic at one spot. Thread the elastic through one lid and tie the weight to the two bands.
4. Put the lid, bands, and weight on one end of the can. Thread the elastics through the other lid, pull the elastic snugly, and place the second lid on the can. Tie the elastics.
5. Roll the tin can toy away from you and watch what happens. Explain the can's movement using what you know of elastic energy.

____

____

____

____

____

____

____

Permission granted to reproduce this page for purchaser's class use only.
Copyright © 2000 Trifolium Books Inc.

Name: ____________________

Build a Paddle Boat

Use the directions that follow to build a paddle boat.

What You Need

- 1 plastic 2 L bottle
- masking tape
- plastic ice cream container lid
- scissors
- elastic bands
- chopsticks

What to Do

1. Tape one chopstick on each side of the bottle. Make sure about 7 or 8 cm sticks out behind the bottle.
2. Cut four rectangles from the ice cream container lid. Each should be around 8 cm x 5 cm.
3. Fold the rectangles in half lengthwise and tape them together in an X-shape to make your paddle.
4. Slip an elastic band over the ends of the sticks.
5. Position the wheel in the middle of the elastic band and wind it up.
6. Let your boat go in a large sink or tub of water and see what happens!
7. How could you improve your boat design? Can you make it go faster? Can you make it go farther? Can you make it go backwards? Test your ideas and revise your design if necessary. Rewrite the instructions for this activity for your new design. Explain your design goal. For example, your new instructions might be called: Build a Fast Paddle Boat.

____

____

____

____

____

____

____

____

____

Permission granted to reproduce this page for purchaser's class use only.
Copyright © 2000 Trifolium Books Inc.

You might ask students to draw plans for a spring- or elastic-powered vehicle. Plans should include details about materials and safe use of the invention.

During the Walk

This TECHNOWALK is perhaps the most difficult, as it is not easy to find examples of elastic and spring energy in use. However, given the challenge, students will be able to seek out situations that demonstrate these forms of energy. We recommend that you walk outside briefly even though you will not see many examples, and then return for an inside walk to examine the operation of springs in many common school or office supplies. On the way outside for the walk, stop by one of the school's fire doors and point out the spring. Ask students to describe the purpose of the spring (to be sure the door closes in a fire). Possible observations of spring and elastic energy in and around the school include: clothespins, hole punch, hand stapler, staple remover, bull-dog spring clips, and quick release bicycle seats. Although it is not obvious at first sight that elastic and spring energy is involved, it can be seen that the expansion and contraction of the elastic and springs do cause motion.

If you can walk to a sporting goods store, have students observe springs on skis, snowboards, bicycles, and bungee cords.

After the Walk

It is likely that students will see very few, if any, examples of elastic or spring energy at work in the community around them. This is because of the limits of the materials themselves. The energy contained in the materials can only be used in small objects, not in power plants! Students could have a discussion about why springs and elastics have limited use and are therefore not seen in many places.

Extension Activities

Tin Can Toys

Use the instructions on the *Technotalk* worksheet Tin Can Toys to get students to build a toy that uses elastic energy. Have students explain why the can rolls away and then comes back. Encourage students to explore the effects of using heavier elastics, or more than one elastic for their toy. Even younger students can do this activity, although you might want to poke holes in the coffee can lids prior to class.

Build a Paddle Boat

Allow older students to build paddle boats using instructions on thc *Technotalk* worksheet Build a Paddle Boat. Encourage them to use the instructions as a "basic recipe" that they can modify in any way they like. Suggest that they define their design criteria prior to their modification tests.

9 ELECTRIC AND SOLAR ENERGY

TECHNOWALK

Before the Walk

Electric energy is very versatile. Electricity is used to run light bulbs and motors, to heat appliances, and in computers and cables, to communicate. Batteries use stored electric energy.

Solar energy is energy that comes from the Sun. Solar energy gives us light, keeps us warm and provides the energy plants need to grow. Plants provide humans with the food they need to live. People either eat plants that have the Sun's energy stored inside, or people eat animals that have eaten plants. Ultimately, all human energy comes from the Sun. Solar energy also causes wind. Heat from the Sun warms air unevenly around Earth; as air heats, it begins to move. Even the energy in gasoline can be traced back to the Sun. Fossil fuels were formed from decayed organisms that were compressed for thousands of years in the ground.

Solar energy can be focused to generate heat or with solar cells it can be converted into electricity. A number of cells must be joined together to generate useful amounts of electricity. A Saskatchewan firm, Solar Freedom, has created ovens that use reflectors to focus the Sun's rays to cook food. The ovens are ideal for use where fuel is scarce but sunshine is plentiful. With the beach-side version of this oven, pizzas can be cooked in thirty minutes.

plants

calculators

traffic lights

electric wires

appliances

What uses of electricity can be seen from the classroom?

What uses of electricity are essential and which are replaceable with some other form of energy?

In what ways does the Sun provide energy?

How is electric energy used?

Where to Walk
Outside: neighbourhood
Inside: school

How Long to Walk
Outside: 1 hour
Inside: 1 hour

Technotalk

Name: ____________________

Evidence of Electricity and Solar Energy

Record your observations of electric and solar energy at work. Make a check in the column you think most describes each use of energy. If you need more room for answers, draw another chart on the back of this page.

Observation	Essential	Replaceable

Make your sketch of electricity at work below:

Permission granted to reproduce this page for purchaser's class use only.
Copyright © 2000 Trifolium Books Inc.

Technotalk

Name: ____________________

Solar Energy at Work

Draw a picture showing how many ways the Sun provides energy to Earth.

Permission granted to reproduce this page for purchaser's class use only.
Copyright © 2000 Trifolium Books Inc.

Caution students never to look directly at the Sun.

During the Walk

1. On the TECHNOWALK, have students find evidence that electric and solar energy is at work. If necessary, direct student attention to the wires that carry the electricity to stores, houses, homes, and factories.
2. You may have trouble observing solar cells in use while on your walk. Encourage students to use all their senses to observe solar energy at work. Students might observe growing plants, animals that eat plants, warmth on their skin, melting ice or snow, an evaporating pool of water, and so on.

Technotalk Worksheet

1. Have students record their observations on the *Technotalk* worksheet Evidence of Electricity and Solar Energy. Tell students to categorize each observation by placing a check mark in the appropriate column.
2. Standing in view of a few buildings (ideally some commercial with electric signs), ask students to sketch the buildings and all the things that use electricity in view.

Older students might be challenged to calculate how much electric energy they use playing video games. They can find the amount of energy consumed by their television and game consoles on the units. The information is normally listed in watts per hour. Add those two numbers together. Convert the number of watts per hour to kilowatts per hour by dividing by 1000. Provide students with a price per kilowatt hour from your electric bill or from the utility company. Multiply the price by the number students calculated to find out how much it costs to play a game for an hour.

After the Walk

1. Ask students to describe all the examples of electric energy they observed. Have students explain the categories they chose for each example. What would students replace the "replaceable" uses with? Could any "essential" examples be moved to "replaceable" with some inconvenience?
2. Ask students to re-draw the same buildings they sketched on their walk, but imagining that electricity had never been invented. What new structures and mechanisms might replace those that require electricity in their drawing?
3. Evaluate students' understanding of the use of electric energy by asking:
   - What are the advantages of using electric energy?
   - Give examples of how you use electric or solar energy.
   - How can people conserve their use of electric energy?

Extension Activities

Solar Energy at Work

On the *Technotalk* worksheet Solar Energy at Work, have students draw an outdoor scene showing how energy from the Sun affects Earth. Suggest that students include as labels: warmth, plants, animals, humans, wind, light, gasoline, solar cells.

CHEMICAL ENERGY

Before the Walk

Light a match for the class and ask students to describe what kind of energy you are demonstrating. Chemical energy is released when chemicals react with other chemicals. When you light a match, you start a reaction with the chemicals on the end of the match. A fire releases chemical energy stored in the wooden or cardboard match-stick. The energy released in a chemical reaction is transformed into heat, light, sound, or movement. Most chemical energy comes from the ground in the form of fossil fuels such as gas, oil, or coal. You might discuss with students where the energy for fossil fuels originally came from (the Sun).

One of the advantages of chemical energy, at the moment, is that it is relatively inexpensive to burn fossil fuels. One of the problems is that all these forms of chemical energy produce carbon monoxide, carbon dioxide and nitrogen oxide. In terms of pollution, coal is the most noxious. Natural gasoline is the cleanest burning fuel.

Fossil fuels are a non-renewable source of energy. Once they are burned, they can never be used again. This is why technology that uses fossil fuels efficiently is very important. We are working on ways of using renewable resources in the future when non-renewable sources run out. Conservation of chemical energy comes largely from cutting back on the use of mechanisms powered by chemicals, as in walking or riding a bike instead of driving a car.

fire

matches

gasoline

fireworks

What are examples of chemical energy at work?

What are examples of renewable energy sources?

Where does the energy in fossil fuels come from?

What are some problems of using chemical energy?

Where to Walk

Outside: local secondary school to visit a chemistry lab, local gas station, neighbourhood intersection

Inside: boiler room in the school

How Long to Walk

Outside: 1 hour

Inside: 1/2 hour

Name: ____________________

Car Tally

Count the number of vehicles passing through an intersection at several points during the day.

Time of day	cars	trucks	buses	bikes

Permission granted to reproduce this page for purchaser's class use only.
Copyright © 2000 Trifolium Books Inc.

Technotalk

Name: ____________________

Energy Resources

If necessary, use an extra page to record your information.

Research Record

Type of energy: ________________________________

Renewable or non-renewable: ________________________

What is this form of energy used to do?

____

____

____

What are advantages to this form of energy?

____

____

____

What are disadvantages?

____

____

____

I found this information in:

____

____

Permission granted to reproduce this page for purchaser's class use only.
Copyright © 2000 Trifolium Books Inc.

Be sure students can observe traffic flow at a safe distance from the intersection.

What are the advantages of using chemical energy?

What would happen if we ran out of chemical energy?

How can we conserve our use of chemical energy?

Conclude your Mechanisms and Energy unit with the assessment task on page 89 of the Appendix.

During the Walk

1. The use of chemical energy in vehicles will likely be the most common observation. You might walk to a place where students can observe a busy intersection several different times during a day or week to keep track of traffic flow. The *Technotalk* worksheet Car Tally will help students record their results.
2. Students can walk to a gas station to compare prices for different types of fuel, such as regular, plus, and super. They should understand that a more refined gas produces less pollution.
3. On the TECHNOWALK, help students observe and keep a tally of chemical energy at work. If necessary, point out people using chemical energy:
   - for transportation: cars, trucks, motorbikes, buses, trains, power boats
   - for work: motors to drive machinery, lawnmowers, chain saws
   - to produce power: batteries, electric generators
   - to produce heat: gas and oil furnaces, wood stoves, barbecues
   - for entertainment: fireworks, campfires.

After the Walk

1. Use the Car Tally worksheet to discuss the advantages of chemical energy use (get to work fast), and disadvantages (pollution at peak commuter times).
2. Ask students to describe how chemical energy helps them everyday: at home, in school, on their way to and from school.

Extension Activity

Energy Resources

Ask students to research either a renewable form of energy (wind, biomass, solar, water, geothermal) or a non-renewable form of energy. Students can use the *Technotalk* worksheet Energy Resources to keep their research on track.

Careers in Energy

Ask an architect to visit your class to talk about energy efficient building innovations.

Appendix

Contents

Detailed Outcomes

Inquiry, problem-solving, decision-making, recording results, data management including collecting, organizing, displaying (graphing), and interpreting data from a variety of sources, design and building using common materials, as well as community exploration are common to all of the TECHNOWALKS.

Technowalk	Mathematics	Science	Technology
1: Inclined Planes	• measurement • classification	• observation • measurement • testing variables	• problem solving
2: Levers	• measurement • estimation • classification	• observation • drawing	• problem solving
3: Wheels and Axles	• geometry • symmetry	• observation • recording results • friction • control and test variables • rotary motion	• identify mechanical parts of objects • design process • construction
4: Gears	• scale drawings • circumference	• communicate results of investigations	• design • construction of models
5: Pulleys	• ratio • measurement	• rotary motion • function of mechanisms • qualitative observation	• design • use of materials • problem solving

Technowalk	Mathematics	Science	Technology
6: Human Energy	• measurement • time • distance	• types of energy use in the home and community • energy conversion • categorizing	• technology past and present • problem solving
7: Air and Water Energy	• measurement distance • time • scale drawings	• evidence and inference • erosion • energy conversions • kinetic and potential energy	• identifying wind and moving water as renewable energy sources • hydraulics • design • construction
8: Elastic and Spring Energy	• measurement • estimation	• writing and following instructions • potential and kinetic energy • setting design criteria	• design • construction • safety • human needs
9: Electric and Solar Energy	• scale drawings • symbolic drawings • multiplication	• energy conservation • observation	• safety • uses of technology
10: Chemical Energy	• surveying • counting	• identifying and observing chemical changes • pollution • chemical reactions • renewable and non-renewable resources	• benefits and drawbacks of technology • careers in technology

Glossary

Bevel gear – two wheels that intermesh at an angle

Block and tackle – a combination of fixed and movable pulleys

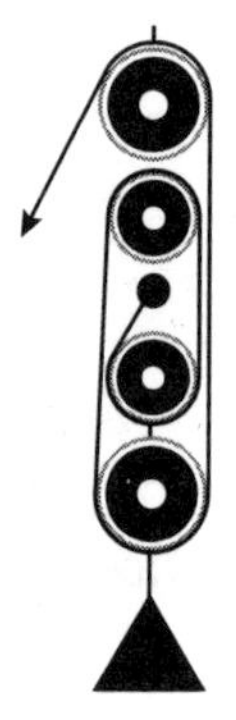

Cog – tooth on a gear wheel

Conservation – saving

Energy – the ability to do work

Friction – a force that occurs whenever two surfaces rub against each other

Fulcrum – the point of rotation of a lever. The fulcrum is also called a pivot.

Gears – wheels with teeth on them

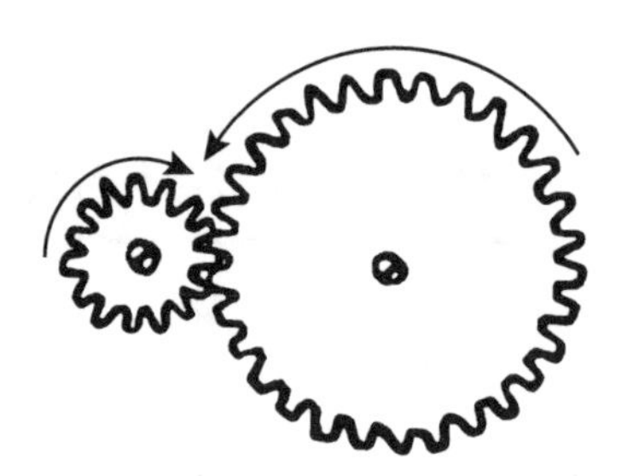

Hydraulics – the controlled movement of liquids from one place to another

Inclined plane – a sloping surface or ramp

Kinetic energy – moving energy

Lever – a rigid rod that rotates around a fixed point

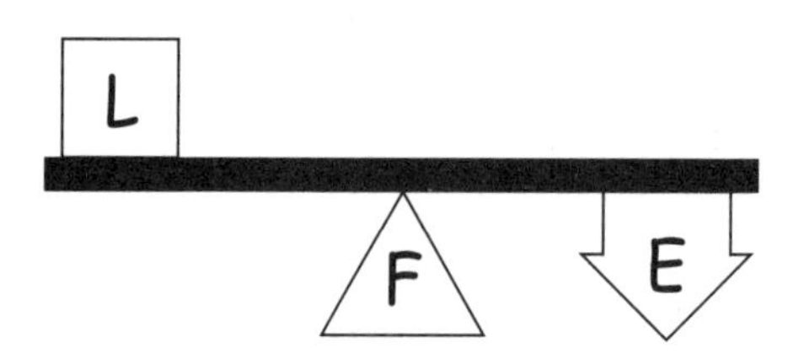

Linear motion – motion in a straight line

Machine – a mechanism that helps people do work

Mechanism – something that uses or creates motion and consists of one or more simple machines

Non-renewable energy – an energy source that cannot be used again

Potential energy – stored energy that can be converted into kinetic energy; energy that comes from position or condition

Power – the rate of doing work

Pulley – a wheel that has a rope or belt around it

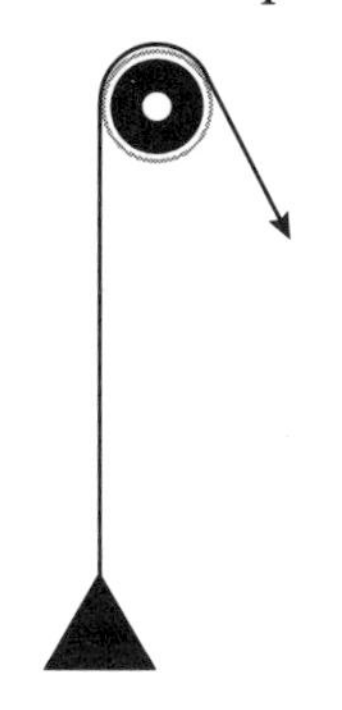

Rack and pinion – a gear working with a sliding toothed rack that converts rotary motion to reciprocating motion

Ramp – a kind of inclined plane

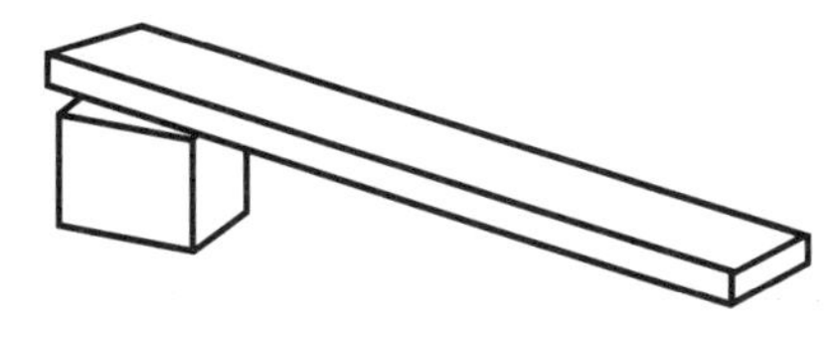

Reciprocating motion – to-and-fro motion

Renewable energy – an energy source that is always being produced

Rotary motion – circular motion around a fixed point

Screw – an inclined plane wrapped around a cylinder

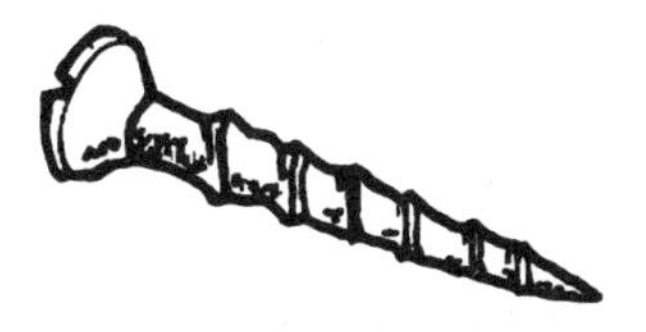

Spur gear – an arrangement of large and smaller gears

Structure – an arrangement of materials

Wedge – an inclined plane that moves

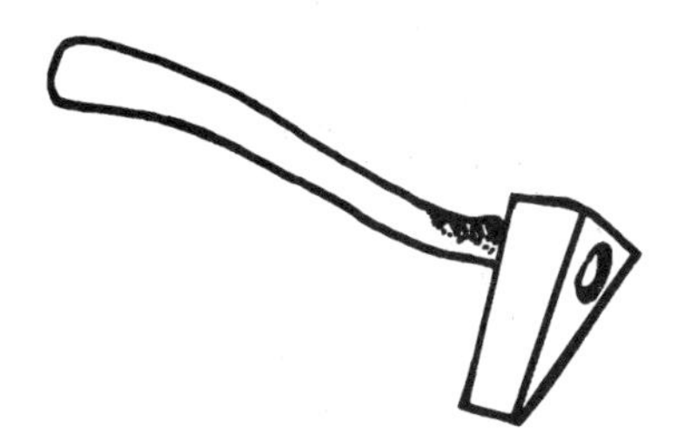

Wheel and axle – a rotating system

Work – the force applied to an object over a distance

Worm gear – a shaft with a screw thread that meshes with a toothed wheel

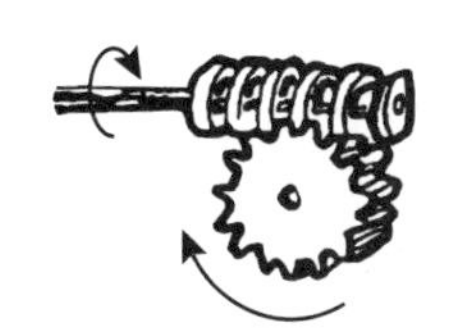

Technotalk

Name: ____________________

What does technology mean to you?

What does the word technology mean to you? In the wheel below, write or draw the first four things that come into your head when you hear the word ***technology***. Exchange your wheel with someone else in your class. Compare your reactions.

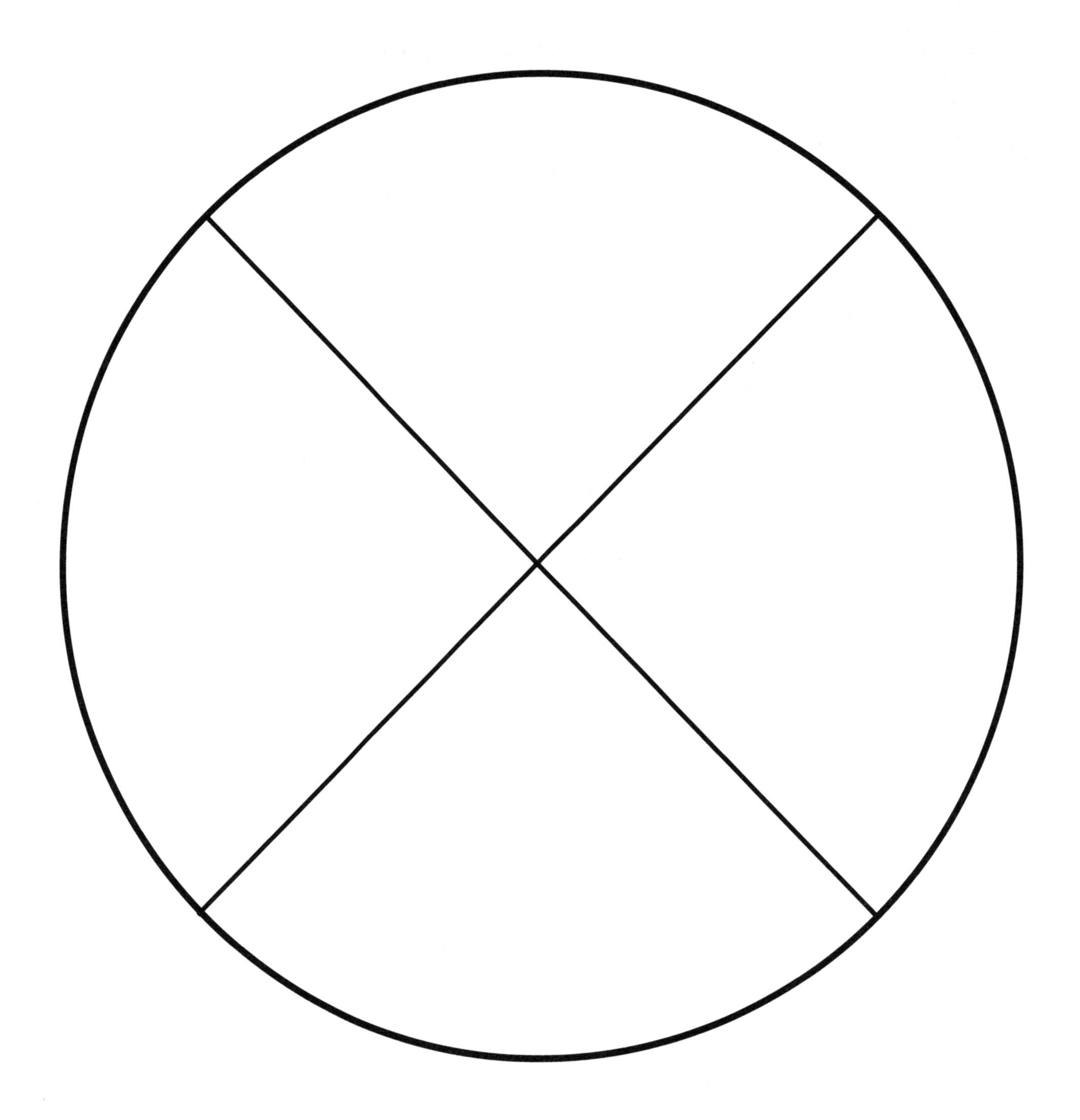

Permission granted to reproduce this page for purchaser's class use only.
Copyright © 2000 Trifolium Books Inc.

Technotalk

Name: ________________

Self-Assessment Form (K-3)

Permission granted to reproduce this page for purchaser's class use only.
Copyright © 2000 Trifolium Books Inc.

Technotalk

Name: ____________________

Self-assessment Form (4-8)

Activity: ____

Circle the number that best describes your work. Think of it as a way of remembering how the project went and what you might do differently next time.

5	excellent
4	very good
3	good
2	satisfactory
1	poor

Effort					
I tried as hard as I could.	1	2	3	4	5
I looked for help when I needed it.	1	2	3	4	5
I helped others in my group.	1	2	3	4	5
Investigating					
I used books and other references.	1	2	3	4	5
I talked to people.	1	2	3	4	5
I used a computer search.	1	2	3	4	5
Construction					
I contributed to the work.	1	2	3	4	5
I listened to others.	1	2	3	4	5
I had a task of my own.	1	2	3	4	5
Time					
I did not waste my time.	1	2	3	4	5
I did not waste anyone else's time.	1	2	3	4	5
I finished on time.	1	2	3	4	5

Choose one item from the ones above in which you think you could improve. What would you do differently next time? ____

____

Permission granted to reproduce this page for purchaser's class use only.
Copyright © 2000 Trifolium Books Inc.

Technotalk

Name: ____________________

Peer Assessment Form (4-8)

Activity: ____

Name of student being assessed: ______________________________

Circle the number that best describes the student's work on the activity just completed. Try to be fair and honest.

5	excellent
4	very good
3	good
2	satisfactory
1	poor

Effort					
Concentrated on the task.	1	2	3	4	5
Looked for help when it was needed.	1	2	3	4	5
Helped others in the group.	1	2	3	4	5
Investigating					
Used books and other references.	1	2	3	4	5
Contacted people.	1	2	3	4	5
Used a computer search.	1	2	3	4	5
Construction					
Contributed to the work.	1	2	3	4	5
Listened to others.	1	2	3	4	5
Took responsibility for a task.	1	2	3	4	5
Time					
Did not waste my time.	1	2	3	4	5
Did not waste anyone else's time.	1	2	3	4	5
Finished on time.	1	2	3	4	5

Choose one item from the ones above in which you think this student could improve. Suggest what the student might do differently next time. ______________________

____

Permission granted to reproduce this page for purchaser's class use only.
Copyright © 2000 Trifolium Books Inc.

Technotalk

Name: ______________

Group Assessment Form (4-8)

Activity: ______________________________

Team members: ______________________________

Agree on your answers as a group. Circle your choice.

1. Did we share?	yes	sometimes	no
2. Did we take turns?	yes	sometimes	no
3. Did everyone contribute?	yes	sometimes	no
4. Did we listen to each other?	yes	sometimes	no
5. Did we help one another?	yes	sometimes	no

Finish each sentence.

6. We agreed on ______________________________

____

7. We disagreed on ______________________________

____

8. We each had a task. The tasks were:

____

____

____

____

9. We could improve by

____

Permission granted to reproduce this page for purchaser's class use only.
Copyright © 2000 Trifolium Books Inc.

Student Name: ____________________

Date: ____________________

Activity: ____________________

Assessment (K-3)

Observed Criteria	Ranking (1 poor to 5 excellent)
confidence	
imagination	
practical skills	
organization	
leadership	
manual dexterity	
positive attitude	
math skills	
creativity	
flexibility	
reasoning ability	
co-operation	
planning ability	
motivation	
language skills	
hand-eye co-ordination	

Permission granted to reproduce this page for purchaser's class use only.
Copyright © 2000 Trifolium Books Inc.

Assessment (4-8)

Student Name: ____________________

Date: ____________________

Activity: ____________________

Observed Criteria	Ranking (1 poor to 5 excellent)

Permission granted to reproduce this page for purchaser's class use only.
Copyright © 2000 Trifolium Books Inc.

Reading List: Mechanisms

Ages 6-8

Armitage, Ronda and David. *The Lighthouse Keeper's Lunch.* London: Andre Deutsch Ltd., 1977. (Every day, Mr. Grinling, the lighthouse keeper, has his lunch sent to him by an ingenious pulley system. One day, however, hungry seagulls eat his lunch and Mrs. Grinling had to come up with a way of outwitting the greedy gulls.)

Armitage, Ronda and David. *The Lighthouse Keeper's Rescue.* London: Andre Deutsch Ltd., 1989. (Two other entertaining stories about the lighthouse keeper and his wife and the use of maritime technology.)

Burton, Virginia Lee. *Mike Mulligan and His Steam Shovel.* Boston: Houghton Mifflin, 1939. (The story of an Irish steam shovel man and his steam shovel — the Mary Anne. A wonderful look the way a complex piece of machinery works.)

Burton, Virginia Lee. *Katy and the Big Snow.* Boston: Houghton Mifflin, 1943. (The story of Katy the crawler tractor, who is a bulldozer in summer and a snowplow in winter. An excellent story about the big machines in our lives.)

Hutchins, Hazel Jean. *Leanna Builds a Genie Trap.* Scarborough: Annick Press Ltd., 1986. (Leanna loses her favourite block. She believes that a genie took it and designs and builds a number of traps out of household materials to catch it.)

Steig, William. *Doctor De Soto.* New York: Farrar, Strauss and Giroux, 1982. (Doctor De Soto is a mouse who is a dentist. He treats animals large and small. For extra-large animals he has a room fitted with pulley systems. There are wonderful illustrations of the machines used by the doctor and his wife.)

Strickland, Paul. *Machines Are as Big as Monsters.* Toronto: Kids Can Press, 1989. (The machines in this book are like gigantic monsters from another world. There are land machines, sea machines, flying machines, and even space machines — and all of them are real.)

Ages 9-14

Asimov, Isaac. *I, Robot.* Garden City, New York: Doubleday & Company Inc., 1950. (A series of nine related short stories chronicling robot development from its crude beginnings in the mid-twentieth century to such a state of perfection a hundred years later that the robots are running the world.)

Green, John F. *Junk-Pile Jennifer.* Richmond Hill, Ontario: North Winds Press, 1991. (Jennifer loves to collect junk. One day she has a big crash with her comic and television hero Captain Astroblast and Jennifer uses up her junk to rebuild the captain's spaceship.)

Hughes, Ted. *The Iron Man.* London: Faber & Faber. (After a clanking iron giant topples from a cliff and smashes, his various parts get up and search for each other.)

Simon, Seymour. *Einstein Anderson.* New York: The Viking Press, 1980. (A series of books about Adam Anderson, a sixth-grader who is such a whiz at science that everyone calls him Einstein. He can solve any mystery or unravel any puzzle simply by applying sound principles of science and technology.)

Information/Activity Books

Baker, Wendy & Haslam, Andrew. *Machines.* Richmond Hill, Ontario: Scholastic, 1994.

Lafferty, Peter. *The Big Book of How Things Work.* New York: Gallery Books, 1990.

Lambert, Mark & Hamilton-MacLaren, Alistair. *Machines.* Hove, England: Wayland, 1991.

Lambert, Mark. *Technology.* Hove, England: Wayland, 1991.

Macaulay, David. *The Way Things Work.* Boston: Houghton Mifflin Company, 1988.

Seller, Mick. *Wheels, Pulleys & Levers.* New York: Gloucester Press, 1993.

Reading List: Energy

Ages 6-8

Browne, Eileen. *Tick-Tock.* Cambridge, Massachusetts: Candlewick Press, 1995. (While Skip's mother is out, this lively squirrel and her friend accidentally break the cuckoo clock. The friends take the clock to be fixed: first to a bicycle mender, then a shoe mender, and finally to a clock mender. This story has a surprise ending — especially for Skip's mother.)

Lovic, Craig. *Andy and the Tire.* New York: Scholastic, 1987. (Andy is at a new school but, apart from a hampster, he has no friends. He finds a tire and uses his own power to travel around to try make himself popular with the children at the school. In the end, the tire makes Andy popular, but not in the way he had imagined.)

Peppe, Rodney. *Mice and the Flying Basket.* Puffin Books, 1994. (A family of mice makes baskets for a living, but they do not bring in enough money. The mice enter a contest called the Flying Circus by turning their baskets into flying machines. The book includes illustrations showing the mice exploring different power sources.)

Spier, Peter. Illustrator. *The Erie Canal.* Garden City, New York: Doubleday & Co. Inc., 1970. (A nineteenth-century journey through the Erie canal with many visual examples of the uses of power and energy.)

Spier, Peter. Illustrator. *Bored — Nothing To Do.* Garden City, New York: Doubleday & Co. Inc., 1978. (Two boys with nothing to do decide to build an airplane from all sorts of materials found around the house. They manage to get it off the ground and fly it before it crashes and their parents find out.)

Ages 9-14

Baisch, Chris. Translated by Mernan, Andrea. *When the Lights Went Out.* Putnam, 1986.

Bourne, Miriam Anne. *Bright Lights to See By.* Coward, McCann & Geoghan, 1975.

Cleary, Beverly. *Dear Mr. Henshaw.* New York: Morrow, 1983. (Part of this story involves Leigh Potts, a sixth-grade student, building an electric burglar alarm to stop people from stealing his lunch from his lunch box.)

Simon, Seymour. *Einstein Anderson.* New York: The Viking Press, various dates. (A series of books about Adam Anderson, a sixth-grader who is such a whiz at science that everyone calls him Einstein. He can solve any mystery or unravel any puzzle simply by applying sound principles of science and technology.)

Information/Activity Books

Cobb, Vicki. *Chemical Active! Experiments You Can Do at Home.* Harper Jr, 1985.

Markle, Sandra. *Power Up: Experiments, Puzzles, and Games Exploring Electricity*. Atheneum, 1989.

Rising, Trudy and Williams, Peter. *Light Magic and Other Science Activities about Energy.* Toronto: Greey de Pencier Books, 1994.

Links with the Community

Who can help you? A list of contacts can be one of your most useful resources.

Contact Name	Date Contacted	Role in the Class	Phone Number/Address

Permission granted to reproduce this page for purchaser's class use only.
Copyright © 2000 Trifolium Books Inc.

What is it made of?

How is it built?

What makes it go?

What is it used for?

How does it work?

How does it look?

Who helps make it?

Dear Parents/Guardians:

The students in our class will be going out on TECHNOWALKS. These are walks in the community to explore technology and how it affects people and the environment. We will begin with an exploration of mechanisms (how things work) and then look at energy (what makes things go).

Aspects of technology will be introduced in the classroom prior to the actual TECHNOWALK and each walk will be followed up with activities back in the classroom and at home. We encourage you to share in this learning experience by discussing with your child the different activities we have been doing. We would also like to invite you to visit our classroom or to join us on our TECHNOWALKS at any time.

Thank you.

Please return the form below with your student.

••

_ _ _ _ _ _ _ _ _ _ _ _ _ has my permission to leave school property for TECHNOWALK excursions with the class. I understand that such excursions involve risks and situations beyond the regular functioning of the school. I will remind my child of the importance of safe, responsible behavior during such excursions.

____

Signature of Parent/Guardian

____

Date

Permission granted to reproduce this page for purchaser's class use only.
Copyright © 1997 Trifolium Books Inc.

Technotalk

Name: ____________________

Reviewing Mechanisms and Energy

Answer each question in the space provided. You may use a web, drawing, chart, words, or any other way of showing how much you know.

How do things work?

What makes things go?

Permission granted to reproduce this page for purchaser's class use only.
Copyright © 2000 Trifolium Books Inc.

Teacher Resources

Print Materials

Chapman, C., et al. *Collins Technology for Key Stage 3: Design and Technology, the Process.* 1992, London: Collins Educational. (Good ideas to help you with the design process.)

Corney, Bob & Dale, Norman. *Technology I.D.E.A.S.* Canada. 1992. Pearson Education Canada Inc. (Projects for elementary students.)

Richards, Roy. *An Early Start to Technology.* London. 1990. Simon & Schuster. (Emphasis on things children can readily design and make through problem-solving.)

See Page 92 for a detailed listing of other Science and Technology Teacher Resources available from Trifolium Books.

Recommended Construction Kits

Grades K-3

Duplo. (Larger version of Lego designed for young children. Excellent introduction to Lego. Permits imaginative play.)

Lasy (Imaginit). (Plastic rectangles, connectors, etc. Versatile.)

Stickle Bricks. (Squares, triangles, and other pieces that "stick" together. Can be combined with "straws." Easy to use.)

Grades 4-8

Baufix. (Wood and plastic, similar to Meccano. Students use nuts, bolts, and spanners. Durable and colourful.)

All Grades

Lego. (Very versatile. Suggest transferring the pieces in a large box with many small compartments, rather than the original cardboard box. Some components are fragile.)

Meccano. (Original version can be difficult for younger students, but there is a new set with plastic that is easier for them to use.)

Ramagon. (Used by NASA engineers to model space stations. Easy to use and works with other construction kits. Durable.)

Internet Sites

The following Yahoo search page is specifically of interest to students and teachers. It allows users to search for information using keywords, but confines its search to things that might be of interest to educators.

http://beta.yahoo.com/education/k_12

- The following web site, posted by *Discover Magazine*, is full of links to other science and technology related sites. It is an ideal research starting point for students.

http://www.enews.com:80/magazines/discover/

Other Sites of Interest

A learning kit of activities and lesson plans focused on motion.

http://schoolnet2.carleton.ca/english/worldinmotion/index.html

Elementary and Junior High. A young student describes 3 successful projects that could be used as activities. Erosion, Battery Power, Soil Pollution.

http://megamach.portage.net:80/~bgidzak/nick.html

Australia: A girls' science and technology high school has a webpage with links to some of their own experiments and projects in robotics and other subjects.

http://www.ozemail.com.au/~mghslib/projects/mghsproj.html

An American webpage similar to the Australian one above can be found at this address.

http://forum.swarthmore.edu/sum95/projects.html

An express lane for teachers on the Information Highway!

Also published by Trifolium Books

The award-winning

The Teacher's Complete & Easy Guide to the Internet

Now available in its completely revised and updated Second Edition

The Teacher's Complete & Easy Guide to the Internet, Second Edition is an indispensable resource for teachers looking for ways to use the Internet effectively in the classroom, while increasing their own comfort level on the Web. The Second Edition expands on the best-selling, award-winning (Best Education Title — 1997 Small Press Book Awards) First Edition, and offers a wealth of information on browser options, meta-search tools, Web page development, copyright issues, and a wide range of lesson plan suggestions and options.

The book includes a resource-rich CD-ROM, which contains over 1,000 pre-selected Web sites organized according to more than 15 curriculum areas and divided into resources and lesson plans. This valuable CD-ROM makes it easy to browse, print, and access resources through your Web browser. The inclusion of a CD-ROM streamlines the process of finding high-quality, curriculum-related links, making these sites quickly and easily available to the teacher.

1-895579-44-9 • 368 PAGES + CD-ROM 7.5" X 9" • SOFT COVER ILLUSTRATIONS • $39.95

"In an area that is often confusing and/or intimidating to teachers, this well-written resource provides up-to-date explanations of technical terms and clear, easy-to-understand descriptions of processes.... Recommended for use by teachers of Kindergarten to Grade 12."
— The Ontario Curriculum Clearinghouse (July 1999)

"...provides many strategies, hints, project ideas, internet sites, teaching tips, and curriculum links which should be of great assistance to teachers, whether they are just beginning to use the Internet or are already experienced users..."
— Victoria Pennell, *Resource Links* (April 1999)

Other Science and Technology Titles from Trifolium Books

SPRINGBOARDS FOR TEACHING SERIES

INVENTEERING
A Problem-Solving Approach to Teaching Technology

Bob Corney & Norm Dale

An essential "getting started" resource for teachers of **Grades K–8** wanting to provide their students with hands-on technological experiences.

8½" X 11" • 128 PAGES • SOFT COVER ILLUSTRATIONS • ISBN: 1-55244-014-1 $29.95 CAN. • AVAILABLE 2000

IMAGINEERING
A "Yes, We Can!" Sourcebook for Early Technology Experiences

Bill Reynolds, Bob Corney, and Norm Dale

Packed with ideas to stimulate young students' imagination and creativity as they explore the issues and applications of technology. For teachers of **Grades K–3.**

8½" X 11" • 144 PAGES • SOFT COVER ILLUSTRATIONS • ISBN: 1-895579-19-8 $29.95 CAN. • AVAILABLE

ALL ABOARD!
Cross Curricular Design and Technology Strategies and Activities

By Metropolitan Toronto School Board teachers

This teacher-tested resource helps educators integrate design and technology easily and effectively into day-to-day lessons. For teachers of **Grades K–6.**

8½" X 11" • 176 PAGES • SOFT COVER ILLUSTRATIONS • ISBN: 1-895579-86-4 $21.95 CAN. • AVAILABLE

Take a Technowalk to Learn about Materials and Structures

Peter Williams & Saryl Jacobson

Provides teachers of **Grades K–8** with 10 fun Technowalks designed to encourage students to investigate the materials and structures that surround us.

8½" X 11" • 96 PAGES • SOFTCOVER ILLUSTRATIONS • ISBN: 1-895579-76-7 $21.95 CAN. • AVAILABLE

TEACHERS HELPING TEACHERS SERIES

BY DESIGN
Technology Exploration and Integration

By the Metropolitan Toronto School Board Teachers

Over 40 open-ended activities for **Grades 6–9** integrate technology with other subject areas.

8½" X 11" • 176 PAGES • SOFT COVER • ILLUSTRATIONS • ISBN: 1-895579-78-3 • $39.95 CAN. • AVAILABLE

Mathematics, Science, & Technology Connections

By Peel Board of Education Teachers

Twenty-four exciting integrated Math, Science, and Technology activities for **Grades 6–9**.

8½" X 11" • 160 PAGES • SOFT COVER • ILLUSTRATIONS • ISBN: 1-895579-37-6 • $39.95 CAN. • AVAILABLE

NEW FOR 2000 — *"Experimenting with..." series*

Fun and interesting activities introduce Grades 4–8 students to hands-on science. Students will develop analytical skills and creative thinking while learning about their physical world.

Experimenting with Air
By Gordon R. Gore

8½" X 11" • 46 PAGES • SOFT COVER ILLUSTRATIONS • ISBN: 1-55244-042-7 $10.00 CAN. • AVAILABLE

Experimenting with Electricity
By Gordon R. Gore

8½" X 11" • 46 PAGES • SOFT COVER ILLUSTRATIONS • ISBN: 1-55244-040-0 $10.00 CAN. • AVAILABLE

Experimenting with Forces
By Gordon R. Gore

8½" X 11" • 46 PAGES • SOFT COVER ILLUSTRATIONS • ISBN: 1-55244-032-X $10.00 CAN. • AVAILABLE

Experimenting with Energy
By Gordon R. Gore

8½" X 11" • 46 PAGES • SOFT COVER ILLUSTRATIONS • ISBN: 1-55244-044-3 $10.00 CAN. • AVAILABLE

Experimenting with Simple Machines
By Gordon R. Gore

8½" X 11" • 46 PAGES • SOFT COVER ILLUSTRATIONS • ISBN: 1-55244-038-9 $10.00 CAN. • AVAILABLE

Experimenting with Light and Colour
By Gordon R. Gore

8½" X 11" • 46 PAGES • SOFT COVER ILLUSTRATIONS • ISBN: 1-55244-036-2 $10.00 CAN. • AVAILABLE

Experimenting with Physical and Chemical Changes
By Gordon R. Gore

8½" X 11" • 46 PAGES • SOFT COVER ILLUSTRATIONS • ISBN: 1-55244-034-6 $10.00 CAN. • AVAILABLE